Vikram Lunar Lander: Dead or Alive?

VIKRAM LUNAR LANDER:
DEAD OR ALIVE?

Dr. Bharat Thakkar & Mayank Chhaya

Published by Dr. Bharat Thakkar & Mayank Chhaya

Copyright 2019 or Vikram Samvat 2076 by Dr. Bharat Thakkar and Mayank Chhaya

All Rights Reserved

ISBN--**9781703351606**

Printed in the United States of America

Dedication

To those with lifelong commitment to

science, facts and exploration.

Not afraid to fail and not vainglorious in triumph.

Contents

1. Introduction - Pg. 8
2. Prologue – Pg. 14
3. The Moon — Because It's There – Pg. 20
4. ISRO – Pg. 29
5. Here Comes the Moon – Pg. 42
6. Politics of Science – Pg. 55
7. The Science of Falling Objects – Pg. 67
8. Moon & It's Uneven Gravity – Pg. 89
9. Beyond the Vikram Crash – Pg. 95
10. Reliability and Quality Control – Pg. 101
11. Epilogue – Pg. 109
12. Extra Update — Pag. 111

Editor's note about the images used in the book

Most of them are frames grabbed from ISRO and Doordarshan's live broadcast.

Introduction

The Indian Space Research Orgnisation's lunar lander Vikram (Courtesy: ISRO)

There is something unfailingly thrilling about watching space missions, even those that fail and especially those that almost succeed.

In most endeavors in life, there is no price for almost succeeding. However, in scientific missions, particularly those that explore space, failure is often as crucial as success. It is with this philosophy that we set out to offer a quick but substantive primer how and why India's ambitious Chandrayaan 2 mission to the Moon, whose potentially path-breaking component was a soft landing of its lander Vikram and rover Pragyan, failed.

The mission costing about 9.78 billion rupees or approximately US$141 million was designed to demonstrate the Indian Space Research Organisation's (ISRO) technical prowess in soft-landing a lander weighing 1,471 kg and rover weighing 27 kg on the near side of the Moon's south polar region. The mission was launched from the Satish Dhawan Space Centre in Sriharikota, Andhra Pradesh, on July 22, 2019 at 2.43 pm local time (09.13 UTC).

It reached the Moon on August 20 achieving lunar orbit insertion. Vikram along with Pragyan separated on September 2 after five lunar-bound maneuvers. Following two deorbiting burns, lasting 4 seconds and 9 seconds respectively on September 3, Vikram began its power descent on September 6. It was scheduled to have a burn time of 15 minutes, the most crucial period of the mission before its actual landing. It was described by ISRO chief K. Sivan as "15 minutes of terror", as a direct but unacknowledged inspiration from the way a NASA scientist had described the dramatic descent of the Curiosity rover on Mars in August 2012 as "seven minutes of terror." However, that was a minor omission compared to what was to unfold in the last moments of Vikram's descent.

Going by the sheer ratio of attempts to success in terms of lunar soft-landing missions so far, there was a little more than 50% chance of Vikram succeeding. Of the 38 attempts of soft landing on the lunar surface so far, only 20 have been successful. So far only three countries, America, the erstwhile Union of Soviet Socialist Republics (USSR) and China have successfully soft-landed their missions on the lunar surface. China's Chang'e-3 mission that successfully landed its rover on December 14, 2013. India would have been only the fourth country to follow suit.

The planning for Chandrayaan 2 began in on November 12, 2007 when ISRO and the Russian Federal Space Agency (Roscosmos) signed an agreement to work on the mission. As part of the agreement signed during the government of Prime Minister Manmohan Singh, the plan was that ISRO would be responsible for the orbiter and the rover, Roscosmos would be tasked to make the lander. The mission was formally approved by the Indian government on September 18, 2008. It is a measure of the complications that attend all such missions that despite ISRO having finalized the payload, it was postponed until January 2013. It was again postponed until 2016 because Roscosmos could not produce the lander in time and withdrew from the project after its Fobos-Grunt mission to Mars failed. According to media reports, since technological specifications of Chandrayaan-2's lander were the same as the Fobos-Grunt, the latter's failure was portentous and abandoned.

That forced ISRO to develop its own lander which it named Vikram after Dr. Vikram Sarabhai, the father of India's space program and India's most prolific institution builder. The year 2019 marked the 100th birth anniversary of Dr. Sarabhai. He died relatively young at 52 on December 30, 1971. For any person being regarded as the father of something as ambitious as a country's space program would have been a life-defining accomplishment but Dr. Sarabhai was so eclectic in his approach to nation-building that he consciously founded so many great institutions. As listed by the Indian Space Research Organisation (ISRO), which he pioneered, it is extraordinary to say the least.

- Physical Research Laboratory (PRL), Ahmedabad

- Indian Institute of Management (IIM), Ahmedabad

- Community Science Centre, Ahmedabad

- Darpan Academy for Performing Arts, Ahmedabad (along with his wife Mrinalini)

- Vikram Sarabhai Space Centre, Thiruvananthapuram

- Space Applications Centre, Ahmedabad (This institution came into existence after merging six institutions/centers established by Sarabhai)

- Faster Breeder Test Reactor (FBTR), Kalpakkam

- Varaiable Energy Cyclotron Project, Calcutta

- Electronics Corporation of India Limited (ECIL), Hyderabad

- Uranium Corporation of India Limited (UCIL), Jaduguda, Bihar

It was to commemorate the memory of this extraordinary figure in India's modern history that ISRO's Vikram lander also had so much national emotions invested in it. Although as a die-hard scientist Dr. Sarabhai would have taken the lander's failure in his stride, it has pushed ISRO on the defensive. Inevitably, in a country that is such a raucous and vibrant democracy, where resources are always scarce, space missions may seem like luxury as well as a source of political jostling. The presence of India's Prime Minister Narendra Modi in the ISRO control room on the night of the Vikram landing was, while perfectly acceptable, also fraught with some measure of political triumphalism. The mission's failure spared the country what many of Modi's detractors apprehended would become yet another giant photo-op for the prime minister who is astute at coopting such non-political endeavors for his own partisan purpose.

In his defense, Prime Minister Modi struck all the right cords while bolstering the sagging morale of the ISRO scientists in the immediate aftermath of the Vikram failure. In a gesture that became both celebrated and controversial, ISRO chief Sivan, who went to see off the prime minister, seemed so overcome that he literally used Modi's shoulder to cry on for several seconds. While the sight of a top space scientist breaking down publicly on national television on the prime minister's shoulder reminded the nation of how too human all involved are, it also prompted some vocal criticism from some quarters. The essence of the criticism seemed to be that "Scientists don't cry."

This short but tightly focused-book seeks to capture the drama that attended the last 15 minutes of a remarkable mission. In the process it attempts to explain some fairly detailed calculations in what could have gone wrong and what fate Vikram might have met as it crashed — hard-landed in other words — on the lunar surface.

Prologue

DR. VIKRAM SARABHAI

It is perhaps unparalleled in world history that a new democracy emerging from centuries of subjugation should have the scientific ambition and political will to embark on a space program. That country was India, and the man embodying that rather audacious ambition was Dr. Vikram Sarabhai, arguably India's greatest institution builder, in the late 1950s and early 1960s.

In her definitive biography of Dr. Sarabhai (August 12, 1919 – December 30, 1971), journalist and writer Amrita Shah says, "When exactly Vikram came up with the notion of a space program for India is not known. R. G. Rastogi, his former student, claims to have heard him talk prophetically of setting up a rocket-launching program "by 1963" as far back as in the 1950s. Praful Bhavsar, who had taken a leave of absence from PRL (the Physical research Laboratory in Ahmedabad) to do a post-doctorate work at the University of Minnesota, recalls Vikram telling him something similar in 1959, and adding that he would want him to return to India to help."

To think that in barely 12 years after the emergence of a deeply impoverished but free India, one man should think in terms of setting up a space program was quite extraordinary. Shah's 2007 book 'Vikram Sarabhai: A Life' describes in some dramatic details the uncertainties that attended the launch of the Nike-Apache rocket supplied by the National Aeronautics and Space Administration (NASA). From a leak developing in the hydraulic system of the crane moving the rocket, to a student discovering a worker still fiddling with the launcher controls just before the blast-off, India's nascent space program was attended by considerable drama.

"At 6.25 p.m. (on November 21,1963), the rocket streaked away into the gathering dusk. Minutes later, a sodium vapour cloud emerged high above, tinged orange by the setting sun," Shah writes. Dr. Sarabhai sent home a telegram that said, 'Gee whiz wonderful rocket shot." Just one day after the Nike-Apache launch, Dr. Sarabhai spoke of an Indian satellite launch vehicle. Unlike space program elsewhere, where the dominant purpose was militaristic, Dr. Sarabhai had a development-oriented, even education-focused vision. His multi-layered vision included the use of space program for purposes such as weather forecasting.

In an email interview with Mayank Chhaya, Shah talked about Dr. Sarabhai's motivation for his space endeavors, the means through which he had found a way of meeting both his social and scientific goals.

Q: *We are approaching the 66th anniversary of the first rocket launch by India on November 21, 1963. Why do you think India's space program, pioneered by Dr. Vikram Sarabhai has not found a comprehensive chronicle?*

Journalist and writer Amrita Shah

A: Much of modern Indian history remains to be written. That said, some good books have been written on the space program, Gopal Raj's Reach for the Stars: The Evolution of India's Rocket Program, for example.

Q: *In your biography of Dr. Sarabhai, you capture in some detail the excitement of that day and the challenges that the project faced. What do you think motivated Dr. Sarabhai to think of space program barely a decade and half after independence?*

A: Space was an extension of Vikram Sarabhai's scientific interest in studying cosmic rays. Unlike most scientists, Sarabhai was possessed of a vision that transcended scientific discovery. He wanted to use science for social upliftment and in space technology he saw an opportunity to achieve this goal. It is true that space programs were generally initiated for military purposes and the Indian space program struck a pioneering path. Sarabhai believed that as a poor, underdeveloped country, India needed to prioritize literacy and economic progress. One can trace his strong identification with the national interest to his family's deep involvement in the freedom struggle; many members of his family, including his sister Mridula had been jailed and taken great risks. His aunt Ansuya had founded a union of textile laborers. The family's closeness to Mahatma Gandhi may also have had an impact in shaping his commitment.

Space was an extension of Vikram Sarabhai's scientific interest in studying cosmic rays. Unlike most scientists, Sarabhai

> *was possessed of a vision that transcended scientific discovery. He wanted to use science for social upliftment and in space technology he saw an opportunity to achieve this goal.*

Why the Indian government under Nehru and Indira Gandhi supported his idea of a space program is not as clear. Both leaders had great faith in science and its potential role in national development. They may have trusted Homi Bhabha and Sarabhai as specialists who knew what they were doing.

Q: *At the core of his vision was the idea of leapfrogging development using technology in defiance of the more popular approach of incremental steps. How do you explain that sense of defiance?*

Both Bhabha and Sarabhai believed in using the western experience to leapfrog, i.e., make rapid strides. I would say it was one view, and at a time when so many things were beginning and taking shape for the first time in a newly free country, I don't know if it was necessarily radical.

Q: *Would you say that for someone who created an inordinately diverse number of institutions, unlike any other Indian of any generation before or since, that space was Dr. Sarabhai's core calling?*

A: No. Sarabhai had a complex vision in which all his activities were interconnected. The SITE (Satellite Instructional Television Experiment) which involved taking television to remote villages for example, called upon managerial capabilities, software development, feedback etc., – you can see that building a management college, a market research agency, the community science center etc., (all of Sarabhai's initiatives) were all apiece with this vision. Yes, space did have a very special place in his heart, and it was the means through which he had found a way of meeting both his social and scientific goals.

Q: *This is perhaps talking in the abstract, but do you think there would have been a space program in India without Dr. Sarabhai?*

A: Yes. I believe at some point there would have been a space program. It might have gone the disappointing way of many of India's scientific institutions; it might have been more military – a wing of DRDO (Defence Research and Development Organisation) maybe. Sarabhai's vision, his ability to inspire, and to combine the idealistic with the practical, and of course his stature, gave the space program a foundation and a profile. Dr. Kasturi Rangan, former head of ISRO told me that the space program was still realizing Sarabhai's vision 30 years after his death, which testifies to his far sightedness.

Chapter 1

THE MOON-BECAUSE IT'S THERE

Perhaps the pithiest response to a question why a person wants to do something extraordinarily adventurous came from the famous British mountaineer George Mallory (1886-1924).

Mallory, who was part of two expeditions to Mount Everest in 1921 and 1922, was asked by The New York Times in March 1923, "Why did you want to climb Mount Everest?"

He responded, "Because it's there."

Mallory perished in 1924 during another expedition to the mountain.

Wanting to conquer or visit a place against all odds simply "because it is there" is quintessentially human. It is as true of what we humans do on Earth as it is of what we do in space. The nearest space destination for us has been the Moon. We want to go there because it is there.

And because it is there, the Moon has, for thousands of years, cast its spell on the human mind, particularly in terms of world literature. It has been the ultimate metaphor for feminine beauty in South Asia generally and India particularly for as long as the Indian civilization has existed. A crescent moon resides in the matted tresses of Shiv as a tribute to its primacy in human experience. The number of times the Moon has been evoked by India's poets is so numerous that it is a pointless exercise to enumerate it.

From afar, especially from our earthly vantage, it is bewitching in all its phases. However, from up close, as discovered by lunar missions, it is covered in "charcoal-gray, powdery dust and rocky debris", according to NASA. It is that powdery surface that reflects the sunlight in a way that is so arresting for us here on Earth. From resplendent silver to teasingly pink and everything in between, the lunar hues as seen from Earth are arresting.

The Moon is as old as Earth — 4.5 billion years. In fact, as NASA points out, the leading theory of its origin is that a Mars-sized body collided with Earth that long ago and "the resulting debris from both Earth and the impactor accumulated to form our natural satellite." *"The newly formed Moon was in a molten state. Within about 100 million years, most of the global "magma ocean" had crystallized, with less-dense rocks floating upward and eventually forming the lunar crust."*

"The early Moon may have developed an internal dynamo, the mechanism for global magnetic fields for terrestrial planets. Since the ancient time of volcanism, the arid, lifeless Moon has remained nearly unchanged. With too sparse an atmosphere to impede impacts, a steady rain of asteroids, meteoroids, and comets strikes the surface. Over billions of years, the surface has been ground up into fragments ranging from huge boulders to powder," it says.

With next to no atmosphere and gravity that is about 17% of Earth's the Moon is not a particularly friendly place for humans. Its gravity is also mysteriously uneven. For an answer to this mystery, the authors came upon a study reported in MIT News (Massachusetts Institute of Technology News) by Jennifer Chu on May 30, 2013. The report said, "Ever since the first satellites were sent to the moon to scout landing sites for Apollo astronauts, scientists have noticed a peculiar phenomenon: As these probes orbited the Moon, passing over certain craters and impact basins, they periodically veered off course, plummeting toward the lunar surface before pulling back up.

As it turns out, the cause of such bumpy orbits was the moon itself: Over the years, scientists have observed that its gravity is stronger in some regions than others, creating a "lumpy" gravitational field. In particular, a handful of impact basins exhibit unexpectedly strong gravitational pull. Scientists have suspected that the explanation has to do with an excess distribution of mass below the lunar surface, and have dubbed these regions mass concentrations, or "mascons."

Exactly how these mascons came to be has remained a mystery – until now.

Using high-resolution gravity data from NASA's Gravity Recovery and Interior Laboratory (GRAIL) mission, researchers at MIT and Purdue University have mapped the structure of several lunar mascons and found that their gravitational fields resemble a bull's-eye pattern: a center of strong, or positive, gravity surrounded by alternating rings of negative and positive gravity.

To figure out what caused this gravitational pattern, the team created simulations of lunar impacts, along with their geological repercussions in the moon's crust and mantle, over both the short- and long-term. They found that the simulations reproduced the bull's-eye pattern under just one scenario.

When an asteroid crashes into the moon, it sends material flying out, creating a dense band of debris around the crater's perimeter. The impact sends a shockwave through the moon's interior, reverberating within the crust and producing a counterwave that draws dense material from the lunar mantle toward the surface, creating a dense center within the crater. After hundreds of millions of years, the surface cools and relaxes, creating a bull's-eye that matches today's gravitational pattern."

It may not be prudent to speculate whether uneven gravity played any role in Vikram veering off course but as a scientific curiosity, it is an interesting point to keep in sight. Perhaps more about that later.

The year 2019 marks the 60th anniversary of the first lunar flyby carried out by the erstwhile Union of the Soviet Socialist Republic (USSR) on January 4, 1959. Called Luna 1, they followed it up with many other robotic probes. Prompted by a healthy albeit ideologically driven sense of scientific competition NASA responded with its Ranger series of hard-lander probes which sent many images from the lunar surface. It was in 1965 that NASA broadcast live images from the Moon with its Ranger 9. In 1966, the USSR created history by soft-landing its Luna 9 spacecraft on an area called Oceanus Procellarum. To the scientific community relief, the landing area was strong enough to support such a spacecraft landing. That same year in

May, the US landed its robotic spacecraft, Surveyor 1, which sent back television images from the Moon.

As a NASA backgrounder by Paul D. Spudis, Lunar and Planetary Institute, pointed out, *"Later Surveyor missions (five in all), collected physical data on soil properties, including its chemical composition. Analysis of the lunar surface showed that the dark maria had a composition similar to terrestrial basalt, a dark iron-rich lava, while the highlands near the very fresh rayed crater Tycho were lighter in color and strangely enriched in aluminum. This led to an astonishing revelation about the Moon's early history after the first physical samples were later returned to Earth by the Apollo 11 crew."*

The backgrounder added, *"Apollo was the finest hour of America's space program. In just eight years, we had gone from zero human spaceflight capability to landing men on the surface of the moon. From these missions, scientists developed a new view of the origin and evolution of the planets and of life on Earth.*

The 1968 Christmastime flight of Apollo 8 was a milestone – humans left low Earth orbit and reached the moon, circling it for almost a day. For the first time, people gazed on the moon from orbit. They found it desolate and gray, but saw nothing to prevent journeying the final 62 miles to the surface. In May of 1969, Apollo 10 orbited the moon, testing the lunar lander. It was a dress rehearsal for the manned landing to come. Each of the Apollo missions – and the astronauts who remained in the orbiting Command Module during the subsequent landed missions – took hundreds of high-resolution photographs of the moon's surface. Their visual observations added to the burgeoning knowledge of lunar geology.

In a harrowing descent marked by program alarms from an overloaded computer and freezing fuel lines, Neil Armstrong and Buzz Aldrin in Apollo 11 safely landed in Mare Tranquillitatis (Sea of Tranquility) on July 20, 1969. They walked on the moon for over 2 hours, collecting rocks and soil and laying out experiment packages."

The Moon is not just Earth's groupie forever attached to our fortunes. One of the most significant consequences of the presence of the Moon where it is is to moderate Earth's axis. That in turn stabilizes any wobbling on its axis that gives a livable climate. Of the more than 190 moons in our solar system, our Moon is the fifth largest and for us the only one so intertwined with our lives. Being tidally locked we see only one side of the Moon. In order for us to see the so-called far side or the dark side of the Moon, we had to mount missions to it, the first being by the former Soviet Union in 1959.

For ordinary readers it is hard to imagine what Earth would have been like without the Moon but there is a lot of literature available on the subject. One of the main consequences of the presence of the Moon is that at the very least it is partly responsible for creating conditions that led to life. It was the tides in our oceans, caused significantly by the Moon's gravitational tug that churned materials that most likely engendered life in some ways.

> *The Moon is not just Earth's groupie forever attached to our fortunes. One of the most significant consequences of the presence of the Moon where it is is to moderate Earth's axis. That in turn stabilizes any wobbling on its axis that gives a livable climate.*

These two factors alone—a stable climate over long periods of time and tides in our oceans—more than underscore the direct bearing the Moon has on us. Wanting to go there—apart from the simple reason because it is there—is almost primal for us. There is a strange draw that the Moon has exercised on the human mind and imagination for as long as we have been around.

Of all the basic vital statistics about the Moon, perhaps the one that jumps out is the comparison between the two bodies' equatorial diameter. The Moon's is 2,158.8 miles or a little over 3,474 kilometers and Earth's 7,917.5

miles or about 12,742 kilometers. The Moon is about a little less than a fourth in its diameter compared to Earth That is a sizable body to have that close not to exercise very significant influence on us.

If the Moon did indeed come out of a collision between Earth and a Mars-sized body, then it is our offspring. And as is the wont of offspring, it does tug at us.

Here are some quick facts about the Moon that people often do not know or remember.

Average Distance from Earth: 238,855 | 384,400 kilometers

Orbit and Rotation Period: 27.32 Earth Days

Equatorial Radius: 1,079.6 miles | 1.737.5 kilometers

Mass: 0.0123 of Earth's (a bit more than 1 percent)

Gravity: 0.166 of Earth's (If you weigh 100 pounds (45 kilograms) on Earth, you'd weight 16.6 pounds (7.5 kilograms) on the moon)

Temperature Range: -414 to 253 degrees Fahrenheit (-248 to 123 degrees Celsius)

Chapter 2

ISRO

The Chandrayaan 2 vehicle (Image: ISRO)

As space enterprises go, the Indian Space Research Organisation (ISRO) is a remarkable case study in extracting the most value out of the least budget. With a budget of $1.5 billion for the fiscal 2019-2020 and 16,815 employees its comparison with NASA's $21.5 billion budget and 17,336 employees is odious for many reasons but also instructive for some. The

European Space Agency, another leading presence in space exploration founded in 1975, has a budget of about $6.5 billion with employee strength of 2,200 in 2018.China's space program, now regarded as second only to the US, has reportedly a budget of $8 billion. These numbers illustrate how ISRO has been able to stretch its dollar so far and into some ambitious programs.

In terms of their existence ISRO — 50 years old in 2019 — is not that much younger than NASA — 61 years in 2019. Although NASA's precursor, National Advisory Committee for Aeronautics was formed in 1915, it was only with the advent of NASA in July 1958 in the midst of a growing space race with the erstwhile USSR. The fact that India, a new nation recovering from at least centuries of plunder, chose to create a space program barely 12 years after it became independent was a tribute to the vision of its founders, particularly Dr. Vikram Sarabhai.

As space enterprises go, the Indian Space Research Organisation (ISRO) is a remarkable case study in extracting the most value out of the least budget. With a budget of $1.5 billion for the fiscal 2019-

2020 and 16,815 employees its comparison with NASA's $21.5 billion budget and 17,336 employees is odious for many reasons but also instructive for some.

It was gutsy for a new country but an ancient civilization that is India to think in terms of a space program so soon after nearly two centuries of systematic impoverishment inflicted by the British colonizers. It is a measure of the plunder by the British colonial masters that Utsa Patnaik, a well-known economist, has worked out that between 1765 and 1938 siphoned 9.2 trillion pounds out of India, which in today's value adds up to $44.6 trillion calculated at the rate of $4.8 per pound sterling almost throughout the colonial period.

It was from a vantage point like this that India's space program ought to be seen in a country which then had about 400 million people. Successive ISRO leaderships have been acutely aware that in a country with limited resources their program may be seen as an obscene luxury. Over the last five decades, it has learned to manage its relatively shoestring budget for a space enterprise very well.

According to ISRO's official backgrounder out of its 105 spacecraft missions since 1975, 11 have failed. Aryabhatta was the first spacecraft mission on Apr 19, 1975 which succeeded followed by Bhaskara 1 on June 7, 1979, which too succeeded. Of the 75 launch missions since August 10, 1979 eight have failed. The first Satellite Launch Vehicle (SLV) 3E1 with Rohini Technology Payload failed. ISRO has launched 297 satellites for 33 countries May 26, 1999. It was from 2007 that ISRO's commercial arm Antrix (Space) Corporation picked up impressive momentum in launching satellites for countries on a commercial basis. In 12 years since it has launched 293 satellites. It has been particularly prolific in launches since 2016.

> *It was gutsy for a new country but an ancient civilization that is India to think in terms of a space program so soon after nearly two centuries of systematic impoverishment inflicted by the British colonizers.*

This is an impressive record for any space enterprise but particularly so for one that works with such severe resource constraints. However, as it has happened with many Indian scientific endeavors, in terms of breaking new grounds in space exploration and technology

ISRO's overall performance has been middling. NASA still remains a premier presence for any number of reasons, including its huge budget as well as its ability to draw on some of the finest scientific, engineering and technological minds who converge in America from all over the world. ISRO's inherent constraints notwithstanding, its missions have been watched with great interest. ISRO Mars Orbiter Mission (MOM), also known as Mangalyaan, was launched on November 5, 2013. Costing $73 million it was the cheapest mission to Mars. As of the publication of this book its orbiter had completed five years in orbit around Mars sending 2 terabytes of imaging data. It was expected to last another one year. Dr. Ashwin Vasavada, a key member of NASA's Curiosity mission to Mars, described ISRO's Mars mission as "a great achievement."

Although the Moon, both a real celestial body as well as a celestial metaphor, has been deeply embedded in India's rich culture for millennia, as a destination to explore it has been a recent one. ISRO's Chandrayaan 1 mission to the Moon was announced on August 15, 2003 by Prime Minister Atal Bihari Vajpayee. It was India's first deep space mission. It was launched on October 22, 2008 and ended August 28, 2009.

NASA offers an excellent summary about it:

India's Chandrayaan-1 played a crucial role in the discovery of water molecules on the Moon.

Chandrayaan-1 was India's first deep space mission.

Among its suite of instruments, it carried NASA's Moon Minerology Mapper (M3), an imaging spectrometer helped confirm the discovery of water locked in minerals on the Moon.

The orbiter also released an impactor that was deliberately crashed into the Moon, releasing debris that was analyzed by the orbiting spacecraft's science instruments.

Chandrayaan-1, the first Indian deep space mission, was launched to orbit the Moon and to dispatch an impactor to the surface.

Scientific goals included the study of the chemical, mineralogical and photogeologic mapping of the Moon. In addition to the five Indian instruments, the spacecraft carried scientific equipment from the United States, the United Kingdom, Germany, Sweden, and Bulgaria.

Chandrayaan-1 was launched into an initial geostationary transfer orbit of 140 x 14,180 miles (225 × 22,817 kilometers) at a 17.9-degree inclination.

Over a period of 13 days, the apogee of the orbit was increased by five burns of the spacecraft's 99 pound-force (44.9 kilogram-force) liquid engine that successively raised orbit on Oct. 23 (to 23,500 miles or 37,900 kilometers), Oct. 25 (to 46,430 miles or 74,715 kilometers), Oct. 26 (to 102,300 miles or 164,600 kilometers), 29 Oct. 29 (to 166,000 miles or 267,000 kilometers), and Nov. 4 (to 236,100 miles or 380,000 kilometers).

Finally, the probe successfully entered lunar orbit after a burn that began at 11:21 UT Nov. 8 and lasted about 13.5 minutes. Initial lunar orbital parameters were about 4,660 x 310 miles (7,502 × 504 kilometers).

Between lunar orbit insertion Nov. 8 and Nov. 12, Chandrayaan-1's orbit was reduced gradually so that it ended up finally in its operational polar orbit at about 62 miles (100 kilometers) above the lunar surface.

Two days later, at 14:36 UT, Chandrayaan released its 64-pound (29-kilogram) Moon Impact Probe (MIP). The probe fired a small deorbit motor and then went into freefall, sending back readings from its three instruments until it crashed onto the lunar surface at 15:01 UT near the Shackleton Crater at the lunar south pole.

Indian scientists reported that data from Chandra's altitudinal composition explorer (CHASE), which took readings every 4 seconds during its descent, suggested the existence of water in the lunar atmosphere, although the data remains inconclusive absent further verification.

Chandrayaan-1 experienced abnormally high temperatures beginning late November 2008 and for a time, it could only run one scientific instrument at a time.

In May 2009, the spacecraft was delivered to a higher 120-mile (200-kilometer) orbit, apparently in an attempt to keep the temperatures aboard the satellite to tolerable levels.

Chandrayaan-1 also suffered a star sensor failure after nine months of operation in lunar orbit. A backup sensor also failed soon after, rendering inoperable the spacecraft's primary attitude control system. Instead, controllers used a mechanical gyroscope system to maintain proper attitude.

Last contact with Chandrayaan-1 was at 20:00 UT Aug. 28, 2009. This was short of the spacecraft's planned two-year lifetime, although ISRO noted that at least 95% of its mission objectives had been accomplished by then. The most likely cause of the end of the mission was the failure of the power supply due to overheating.

Perhaps Chandrayaan-1's most important finding was related to the question of water on the Moon. In September 2009, scientists published results of data collected by the American M3 instrument which had detected absorption features on the polar regions of the surface of the Moon usually linked to hydroxyl- and/or water-bearing molecules.

This finding was followed in August 2013 by an announcement of evidence of water molecules locked in mineral grains on the surface of the Moon -- magmatic water, or water that originates from deep in the Moon's interior.

Magmatic water had been found in samples returned by Apollo astronauts but not from lunar orbit until the operation of the M3 instrument. Although Cassini, during its flyby of the Moon in August 1999, had detected (using its VIMS instrument) water molecules and hydroxyl.

NASA's Deep Impact-EPOXI mission, which flew by the Moon in June 2009 also returned the same type of data.

By all accounts it was a successful mission and its highlight was the discovery of water molecules on the Moon.

With that impressive first to its name, ISRO had already been working on Chandrayaan 2. It was approved on September 18, 2008 by Prime Minister Manmohan Singh. It was launched on July 22, 2019. An ISRO backgrounder explained the motivation behind Chandrayaan 2's launch thus:

"(The) Moon provides the best linkage to Earth's early history. It offers an undisturbed historical record of the inner Solar system environment. Though there are a few mature models, further explanations were needed to understand the origin of the Moon. Extensive mapping of lunar surface to study variations in lunar surface were essential to trace back the origin and evolution of the Moon. Evidence for water molecules discovered by Chandrayaan-1, required further studies on the extent of water molecule distribution on the surface, below the surface and in the tenuous lunar exosphere to address the origin of water on Moon.

The Lunar South pole is especially interesting because of the lunar surface area that remains in shadow is much larger than that at the North Pole. There could be a possibility of presence of water in permanently shadowed areas around it. In addition, South Pole region has craters that are cold traps and contain a fossil record of the early Solar System.

Our Sun emits a continuous outflowing stream of electrons and protons into the solar system, called the solar wind. The solar wind plasma which has charged particles embedded in the extended magnetic field of the Sun, moves at speeds of a few hundred km per second. It interacts with solar system bodies including Earth and its moon. Since the Earth has a global magnetic field, it obstructs the solar wind plasma and this interaction results in the formation of a magnetic envelope around Earth, called the magnetosphere.

The Earth's magnetosphere is compressed into a region approximately three to four times the Earth radius (~22000 km above the surface) on the side facing the Sun but is stretched into a long tail (geotail) on the opposite side that goes beyond the orbit of Moon. Approximately, once every 29 days, Moon traverses the geotail for about 6 days centered around full moon. Thus Chandrayaan 2 also crosses this geotail and its instruments can study properties of geotail at a few hundred thousand kilometers from Earth.

The CLASS instrument on Chandrayaan 2 is designed to detect direct signatures of elements present in the lunar soil. This is best observed when a solar flare on the Sun provides a rich source of x-rays to illuminate the lunar surface; secondary x-ray emission resulting from this can be detected by CLASS to directly detect the presence of key elements like Na, Ca, Al, Si, Ti and Fe.

While this kind of "flash photography" requires one to await an opportune time for Sun to be active, CLASS in its first few days of observation, could detect charged particles and its intensity variations during its first passage through the geotail during Sept.

Why was the Lunar South Pole targeted for exploration?

(The) Moon provides the best linkage to Earth's early history. It offers an undisturbed historical record of the inner Solar system environment. Though there are a few mature models, further explanations were needed to understand the origin of the Moon. Extensive mapping of lunar surface to study variations in lunar surface were essential to trace back the origin and evolution of the Moon. Evidence for water molecules discovered by Chandrayaan-1, required further studies on the extent of water molecule distribution on the surface, below the surface and in the tenuous lunar exosphere to address the origin of water on Moon.

The Lunar South pole is especially interesting because of the lunar surface area that remains in shadow is much larger than that at the North Pole. There could be a possibility of presence of water in permanently shadowed areas around it. In addition, South Pole region has craters that are cold traps and contain a fossil record of the early Solar System."

Three deep space missions, two to the Moon and one to Mars, in a span of 16 years is not extraordinary but all things considered, including India's own socio-economic challenges and ISRO's own tight budgets, it is a highly commendable performance. The build-up to the launch of Chandrayaan 2 was full of excitement, especially because it was coming on the heels of the first one that found water molecules as well as because it was going to attempt a soft-landing.

The Chandrayaan 2's first launch on July 15, 2019, had to be called because of a technical snag. An official statement said, ""India's second mission to Moon, Chandrayaan-2 onboard GSLVMkIII-M1 has been called off due to a technical snag. A technical snag was observed in launch vehicle system at T-56 minute. As a measure of abundant precaution, Chandrayaan-2 launch has been called off for today."

It was eventually launched a week later. The launch was flawless, setting spacecraft on a seven-week journey to the Moon. The unusually long time compared to a matter of a couple of days was explained by the fact that ISRO's Geosynchronous Satellite Launch Vehicle Mark-III does not have the same amount thrust as Saturn V rockets that powered NASA's Apollo program.

Chandrayaan 2 spent 23 days orbiting Earth even as it increased its altitude on one side of its elliptical orbit. It was not until mid-August that it began preparations to leave Earth's orbit and travel towards the Moon.

Chapter 3

HERE COMES THE MOON

Anxious ISRO scientists tracking the last moments of Vikram's descent.

Space mission control rooms everywhere have a certain air about them—an air of anxious optimism. ISRO's Chandrayaan 2 mission control was also no different. Irrespective of whether a mission being tracked succeeds or fails, there is always great excitement about watching scientists peering at their computer monitors unwaveringly for any and all hints of where the mission is headed.

The control room scientists' body language is a cornucopia of human emotions if one knows how to read them. While the seasoned ones manage to remain inscrutable like professional poker players, most of them betray signs that can be interpreted to know how the mission is going.

Apart from the historic significance of India's Chandrayaan 2 mission, only the fourth country to attempt soft-landing a probe on the Moon and that too on the near side of the Moon's south polar region, the authors also watched it live to see how mission control staff responded emotionally to every moment.

The Chandrayaan 2 lander, Vikram, was set to attempt a soft landing on the night of September 7 in India and afternoon of September 6 in the United States. The location was, as NASA noted, "a small patch of lunar highland smooth plains between Simpelius N and Manzinus C craters."

The culmination of any mission on any surface — be it on the Moon, Mars or an asteroid — comes after years of intense scientific, engineering, materials, management and even political work. Given the nature of space missions in terms of their cost to any country's exchequer, they may all have begun from a scientific standpoint but invariably get cleared politically.

Almost exactly 11 years after the government of Prime Minister Manmohan Singh cleared the project on September 18, 2008, it was coming down to the last 15 minutes of its long journey above the Moon as the Vikram lander and its rover colleague Pragyan began its descent.

Unlike other mission control rooms such as those of NASA's or China National Space Administration's (CNSA), where there is sartorial uniformity, ISRO's was in keeping with the exuberant diversity of the country it so proudly represents. It was a mosaic of colors, especially among the women scientists. While uniformity in terms of everyone wearing the same kind of T-shirts shirts does give a mission control room a certain earnestness of purpose, the eclectic nature ISRO's mission control seemed to work in the Indian context. While NASA, CSNA or European Space Agency's (ESA) control rooms may look straight out of a Hollywood movie, ISRO's tends to be more informal. Not that one is necessarily better than the other. It is just symptomatic of a country's cultural predilections.

There were two broad ways to gauge the progress of the Vikram landing in the last 15 minutes, described by ISRO chief K. Sivan as "Fifteen minutes of terror" drawn from NASA's characterization of its Mars Curiosity rover's landing as "Seven minutes of terror". One was to study the telemetry graph and watch how closely it followed the normal path. The other was to see subtle mood changes on the faces of and the body language of the scientists.

About 12 minutes into live broadcast of the lander's descent the ISRO personnel at the Mission Operation Complex, Telemetry Tracking and Command Network looked anxious but confident. At the countdown point minus 22 minutes for the power descent of the lander the control room staffers began their peer close at their monitors and swift on their keyboards. At that point there were five events lined up — start of descent, rough braking end, fine braking start, fine braking end, vertical descent start, central engine on and touchdown. Just before final descent the spacecraft was orbiting 30 kilometers perilune, or the closest to the Moon, and 100 kilometers apolune or the furthest point from the Moon. It was an elliptical orbit. At countdown minus 19, the plot of the landing descent trajectory began to appear. The authors were paying particular attention to it. At that point Vikarm was set to cover 600 kilometers at touchdown. It was just around that time when Prime Minister Narendra Modi arrived to watch the soft-landing live.

A round of applause among the scientists began at altitude 30.425 km even as the prime minister joined in while the rough braking phase was initiated to last about ten minutes. That was meant to herald the first part of the descent phase. At this point the velocity of the lander 1640 meters per second which had to be gradually dropped to 0 in time for its soft landing. However, that point was still about 15 minutes away. A minute and 28 seconds into the descent, the velocity was reduced to about 1507 meters per second even as faces in the room begin to beam. A group of school students, who were specially invited for the landing, was seen animatedly talking among themselves and pointing at the giant monitoring screen panels.

Five minutes into the descent, the horizontal velocity dropped to 1009 meters per second while vertical velocity was at 34.1 meters per second with a down range of 202 km. At some 990 meters per second horizontal velocity it was announced that the propulsion was going as planned by the computer onboard the lander. One could see the prime minister sit up six minutes into the phase and peering at the screens.

At the altitude of about 20 km as the lander was descending at under 800 meters per second, it was observed that the plot was matching precisely with the pre-flight predictions. That was seen as the clear indication that the four throttleable liquid-fuel steering engines, designed by India, were working perfectly. They fired for 11 minutes, apparently as designed, to complete the lander's "rough braking phase". That guided the craft to an altitude of around 7.4 kilometers or 24,000 feet. At that stage Vikram was lower than the height of Mount Everest of 29,029 feet.

As it approached some 75 kilometers from the landing site it had covered over 525 kilometers when the descent maneuver was initiated. At that point the lander was traveling at 390 meters per second horizontal velocity and 73.9 meters vertical velocity. The engines continued to fire as part of the rough braking phase. As the lander reached the velocity of 268 meters per second, the imager data download function also came alive. Once again, a round of applause went off in the mission control.

The rough braking phase ended soon after that as the screen panels showed the altitude pf 6.111 km. Next came absolute navigation and control phase that would have lasted about 38 seconds. With the velocity of 86 meters per second the plot began to show more rapid downward movement still coinciding with the pre-flight graph. At that stage Vikram was about 4.3 km from the landing site. While watching the live broadcast of the soft-landing of India's Chandrayaan 2 mission from Chicago the authors' eyes were mostly focused on the telemetric graph of the descent trajectory whenever it was shown. For about 90 percent of the time the descent was precisely going as expected.

Then around 2.00 a.m. India time (3.32 pm CST here in Chicago) there was a slight but distinct deviation in the trajectory, causing us to sit up. However, the descent returned to match the projected path for a short while.

The biggest round of applause came with the announcement that the rough braking phase had been successfully completed. That was the beginning of the fine braking phase which would have lasted 96 seconds. That's when it would eventually to reach an altitude of 400 meters.

The velocity had dropped to horizontal velocity of 48 meters per second and vertical velocity of 60 meters per second and the lander was within reach of the lunar surface.

> *Even sitting some 8500 miles or nearly 13,700 kms away from the ISRO control room, one of the authors was beginning to notice a distinct shift in the mood there. One could sense on the faces of the mission control staff that the absence of any more movement on the graph was problematic.*

That earlier wobble though kept nagging us purely out of intuition. Soon afterwards, the descent deviation became dramatic and then froze sending data. At that point, which must be around 3.36 pm CST or 2.04 a.m. IST, that I posted on my Facebook timeline. "Clearly, there is a glitch with Vikram landing," the update said. Eight seconds later, I posted this: "Vikram landing is not looking promising. I hope I am wrong."

Even sitting some 8500 miles or nearly 13,700 kms away from the ISRO control room, one of the authors was beginning to notice a distinct shift in the mood there. One could sense on the faces of the mission control staff that the absence of any more movement on the graph was problematic. It was around that time that Indian Space Research Organisation (ISRO) chief K. Sivan, looking crestfallen, went up to Prime Minister Narendra Modi.

The horizontal and vertical velocities seemed to be frozen at 48 meters and 59 meters per second. It was around 53.06 minutes into the descent broadcast that one of the scientists was seen holding his head in his hands. Taken together, it was becoming clear that the mission had run into a problem.

There is a difference in a silence of hopeful anticipation and a silence of intimations of failure. This silence was the latter. Tensions began to distort some of the faces in the mission control. Postures in chairs began to sag. Thirteen minutes and 48 seconds into the descent, things were not looking good at all. The first slight deviation, which may well have been perfectly normal to the scientists, had caused in us great unease that was reinforced by the second more pronounced wobble in the tracking.

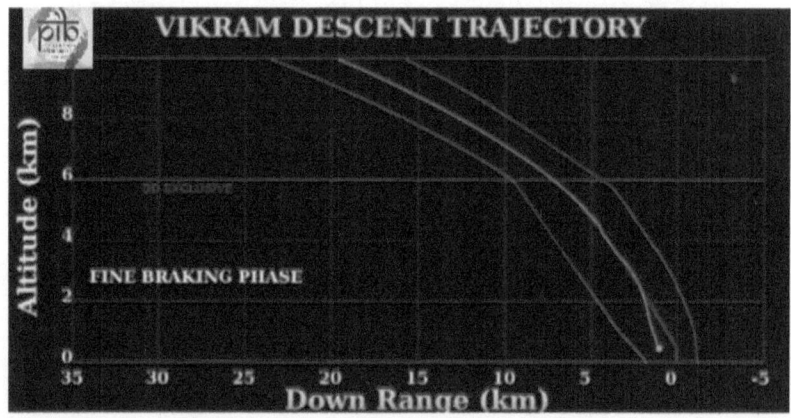

The fatal deviation in the lander's descent

> *There is a difference in a silence of hopeful anticipation and a silence of intimations of failure. This silence was the latter. Tensions began to distort some of the faces in the mission control. Postures in chairs began to sag. Thirteen minutes and 48 seconds into the descent, things were not looking good at all.*

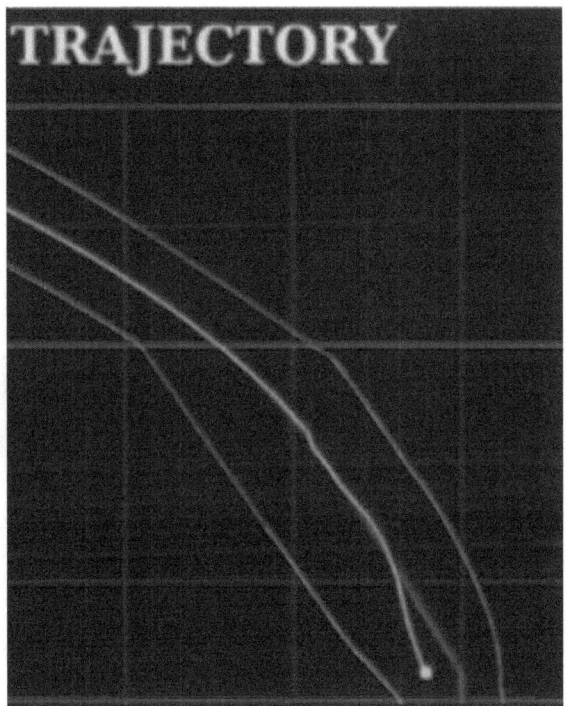

One of the two authors, Mayank Chhaya, although a journalist of longstanding with a degree in physics, has no training at all in these matters. He was going purely by what the telemetry was showing and the deviation and subsequent freezing coupled with just raw intuition. Sitting in his study in a different town some distance away, Dr. Bharat Thakkar, 77, a veteran of quality control and reliability of systems, who has taught at several prestigious institutions here in

Chicago, was already embarked on his calculations. With nearly five decades in studying precisely why systems fail and a proven track record of having predicted many times, Prof. Thakkar was getting the unpleasant sense of what could have gone wrong.

An announcement was heard in the background saying, "We are not receiving from Madrid, HBK and Mauritius."

Over 17 minutes into the descent, a full two minutes later than it should have, it was clear that the lander had not had a successful soft-landing.

With the plot showing no movements at all, the mood began to sink even deeper. It was then that ISRO boss Sivan and immediate colleagues pushed their chairs back, clearly looking defeated. The mood darkened with some scientists clutching their heads in their palms. Hands were clasped and deep breaths were inhaled. Sivan began walking up to Prime Minister Modi.

The huddle with the prime minister was full of briefing in low tones, signaling that Vikram had not landed successfully. Sivan's back was patted by his current former colleagues even as the prime minister settled back into his sofa showing every sign that things had not gone well. Soon afterwards, he went down to the control room from the viewing gallery.

For some strange reason ISRO chief Sivan announced. "Vikram lander was as planned and the normal performance was observed up to an altitude of 2.1 kilometers. Subsequently, communication from the lander to the ground station was lost. The data is being analyzed." He repeated that announcement.

"It is not a small feat we have achieved. Be courageous," the prime minister said and patted Sivan's shoulder with particular vigor. "I am always happy," he said.

"I am proud of you," he said in Hindi. "As many scientists told me, if the communication starts again, it (Vikram) will yield many things. So hope for the best. Many congratulations from me. You have served our country very well. You have served science well and you have served humanity well. We will learn a great deal from this and our journey will continue. I am firmly with you. Go ahead with courage. With your hard work, the nation will celebrate again; I am confident. Wish you all the best," Modi said.

The failure of the Vikram landing is a deeply sobering one for ISRO and no one knows it better than them. There is no point holding forth on the causes unless you have a working grasp on the sciences involved. If you cannot tell the difference between the horizontal velocity and vertical velocity—the former is constant while the latter changes every second beginning at 9.8

meter per second — you have no business reporting such stories.

It is precisely because Dr. Thakkar has spent a lifetime on such matters, have we ventured with this.

Chapter 4

Prime Minister Narendra Modi, extreme left, in a somber mood.

POLITICS OF SCIENCE

There are a couple more general steps before we get into the mathematical specifics about what might have happened to Vikram. The most obvious one was how India's national mood sank within minutes of Vikram's failure. No single visual became more emblematic of that mood than ISRO chief Sivan welling up as he escorted the prime minister's departure from the mission control. He seemed so overcome that Modi felt compelled to console him by literally lending his shoulder to cry on.

Those images were seen throughout India and around the world, mostly to empathetic and sympathetic reactions. However, there was a distinct undercurrent of dismay among some in the scientific community at the sight of a fellow scientist breaking down over a failed space mission. At the human level, the show of emotions might be natural and understandable but there is a popular view of the scientist being stoic in the face of such failures, particularly space missions where the amount of variables is so high and the rate of failure quite dramatic.

As the ISRO chief, it was understandable that Sivan was so emotionally invested in a mission of this scale and for quite some time. Add to that mix the hopes of a nation riding on that mission and then compound it by political capital being spent by the prime minister personally as well as on behalf of his nationalistic Bharatiya Janata Party (BJP).

For Modi and his party, the constant refrain since they first came to power in 2014 and then returned even more robustly in 2019 has been that most preceding governments with the exception of Prime Minister Atal Behari Vajpayee did next to nothing for India's development. The relentless assertion that "in the last 70 years nothing was done in India", which was at least partially responsible for Modi and the BJP winning not once but twice back to back, would, in their egregiously flawed view, have been handsomely vindicated by a successful soft-landing of Vikram.

We make it a point to emphasize in our Introduction how frequently similar missions by other pioneering countries have failed. Going by the sheer ratio of attempts to success in terms of lunar soft-landing missions so far, there was a little more than 50% chance of Vikram succeeding. Of the 38 attempts of soft landing on the lunar surface so far, only 20 have been successful. So far, only three countries, America, the erstwhile Union of Soviet Socialist Republics (USSR) and China have successfully soft-landed their missions on the lunar surface. China's Chang'e-3 mission that successfully landed its rover on December 14, 2013. India would have been only the fourth country to follow suit.

As space missions go, ISRO scientists ought to have known that soft-landing any spacecraft on any surface with markedly different conditions, including terrain and particularly gravity, is a huge challenge. Everything that can and should go wrong with such landing frequently does. It is in the nature of this enterprise. Even accounting for irrational optimism, the ISRO team must have known that the last mile was the decisive one. Sivan himself had described the last phase as "15 minutes of terror."

Notwithstanding the scientific awareness that things can and will go wrong, hope often overrides all such terrible uncertainties. At times, excessively so. As the great 18th century poet Alexander Pope, so famously wrote in his poem 'An Essay on Man (1733-34), "Hope springs eternal in the human breast."

As soon as Vikram failed, irrational hope began to rise not just among ordinary folk but even the ISRO leadership. Alibis and mitigation began to be chased. Even as acutely aware that two of the three key components of the Chandrayaan 2 mission had essentially failed completely, Sivan made a startling claim in the days that followed. He said that the mission was 98 percent successful. He told the media in Bhubaneswar, "We could not establish any communication with the lander (Vikram) yet. The project was developed in two parts — science and technology demonstration. We achieved total success in science objective while in technology demonstration, the success percentage was almost full. That's why the project can be termed as 98 percent successful."

The claim surprised and even shocked many since logically the Chandrayaan-2 mission had three objectives—putting an orbiter around the Moon, something India had already achieved with Chandrayaan-1, soft-landing a lander and then successfully deploying a rover. If two out of three mission goals were not achieved, how was it a 98 percent success?

> *As soon as Vikram failed, irrational hope began to rise not just among ordinary folk but even the ISRO leadership. Alibis and mitigation began to be chased.*

The wire service, IANS, illustrated the incredulity of some fellow scientists in India by quoting two former ISRO officials. "In one week, (success) percentage has increased by three percent. By the unitary method, in another five days, the success rate will become 100 percent," was the sarcastic comment made by a former official, who did not want to be identified, to IANS.

"In two days, the Indian space agency may announce Chandrayaan-2 mission was 100 percent success," another retired ISRO official told IANS.

Perhaps stung by the scorn and derision, ISRO felt the need to make an official statement that said, "The success criteria was defined for each and every phase of the mission and till date 90 to 95 percent of the mission objectives have been accomplished and will continue contribute to lunar science, notwithstanding the loss of communication with the lander."

It was a carefully crafted and perhaps even politically weighed statement. One can only speculate whether ISRO bosses felt obliged of their own volition or felt political pressure to keep their observations in comport with the popular political narrative that India had begun to push scientific and economic frontiers only since 2014. There is absolutely nothing to suggest that publicly or perhaps even privately that anyone from the Modi government, let alone the prime minister himself, tried to subtly lean on to ISRO to not just extenuate the Vikram failure but even claim a 98 percent success.

On the face of it, Modi's own comments within minutes of the realization of failure in the form of a pep talk were unexceptionable and even remarkably sensible. It is perfectly possible that the prime minister made a political calculation in being personally present in the mission control based the potentially handsome rewards of perception. Had Vikram soft-landed successfully, at the very least it would have created for him the optics of a decisive leader helming an invigorated nation even in the middle of the night. We have no personal knowledge that he did indeed make that calculation. Modi is astute enough to know that space missions can fail at the finishing line. Therefore, the argument that he was solely or even at all driven by political optics does not hold much water even though there is a record of claiming some reflected glory on his part.

On balance it appeared — and this is notwithstanding his justifiably vocal detractors who frequently deride and mock him for his attention-grabbing obsessions — that at least in this particular case the motivation for being on the scene was genuinely non-political. At the same time though, no one should be surprised if he were equally mindful within himself that a success here would buttress his and his party's often unseemly assertive claims of putting India on a positive trajectory only since 2014 and for the in-between phase of the Vajpayee government between 1998 and 2004. Everything other phase in India's history since 1947, in their estimation, has been characterized by malicious ineptitudes of the Congress Party and other governments.

It is undeniable that the India surging only since 2014 narrative would have been enormously boosted by Vikram's success but that is not necessarily the overarching reason behind our view. It is possible that the narrative has so seeped into many parts of the country and particularly high institutions such as ISRO that their grandees reflexively feel the need to live up to it and make claims accordingly in the face of facts.

The two remarkably derisive observations by former unnamed officials by the IANS wire in a sense accentuate this compulsion to be in line with the political ambience. We cannot assert enough that there was nothing in the prime minister's comments that even remotely suggested that at that particular juncture he was pushing that narrative even though he is known to have done so on many earlier occasions.

Scientific research that directly impact ordinary citizens in general and space missions in particular have always had a strongly political dimension to it—be it the development of nuclear weapons in the 1930s and 1940s or the race to "conquer" space between the then Soviet Union and the United States. Nationalistic folklore has always been built around particularly applied sciences. Political parties and their leaders are known to have mined scientific achievements for their political benefits. In a strange sort of way, it even makes sense because when greatly successful, such as the Apollo 11 mission to the Moon of July 1969 for instance, they have a way of rubbing off on the political dispensation of the day. To that extent, it would have been natural for any prime minister, especially someone so photo-op and branding obsessed as Prime Minister Modi, wanting to be seen in the immediate midst of such successes.

However, in the case of the current BJP dispensation as well as its very vocal and often unabashed advocates on social media there is also another narrative that feeds into it. Since coming to power in 2014 acolytes of the BJP have repeatedly spoken about India's ancient glories in scientific and technological achievements in often such outlandish terms that even the truly great ones get subsumed by the over-the-top rhetoric. That an ancient India, and therefore Hindu India in the party's estimation, was so spectacularly successful at science has been a favorite theme. A subset of that has been the argument that the "secular, West-obsessed" government of the first Prime Minister Jawaharlal Nehru never really cared much for India's ancient glories and instead pushed European ideas. Countering that deeply flawed narrative is beyond the scope of this book but suffice it to say that historical reality does not track with this political rhetoric at all.

It is recorded history that India had had great affinity for applied sciences millennia ago. Just one example should illustrate that. The use of nanotechnology has been known to exist among Indian craftsmen and artisans as far back as 2000 years ago. Nobel laureate Robert Kurl Jr., who shared the Chemistry prize with two others in 1996, was quoted as saying at the 95th Indian Science Congress, "Our ancestors have been unwittingly using the technology for over 2,000 years and carbon nano for about 500 years. Carbon nanotechnology is much older than carbon nanoscience." The fashioning of weapons and tools or even color pigments in the frescoes of Ajanta and Ellora and elsewhere that last much longer than normal by unwittingly deploying nano-techniques has been widely known. On the Damascus blades used in Tipu Sultan's sword was found evidence of carbon nanotubes as well as nano particles in the pigments of the frescoes. The technology to produce wootz steel to make high-grade steel has also been recorded in ancient India. The technique mixes wood and other organic material during the forging of the steel. The use of wootz has declined since the 17th century, according to Prof. Curl.

> *The point is that there is enough in India's political ambience lately that consciously or subconsciously pressures even the otherwise rational people to invoke unproven scientific glories. This is tragic because a historical India had enough genuine scientific accomplishment so as not to be debased by absurd exaggerations.*

There are other areas of applied science where India has been known to excel but in the hyperbolic political narrative some of those achievements have become so fanciful as to be ridiculous.

Another more trivial side to this debate has been resort to silly superstitions such as hanging green chilies strung with a thread on new devices, vehicles and instruments to ward off "the evil eye". An accompaniment to that superstition is to keep a lime under the tires of new vehicles that must be crushed before putting it to a regular use. The reason for mentioning this particular superstition is because a personage no less than India's Defense Minister Rajnath Singh performed the ritual of placing two green limes under the wheels of the Rafale fighter jet that India has acquired from France as recently as October 8, 2019. Ironically, it was an old video of Prime Minister Modi mocking precisely such superstitions that went viral parallel to Singh's ritual offering.

The point is that there is enough in India's political ambience lately that consciously or subconsciously pressures even the otherwise rational people to invoke unproven scientific glories. This is tragic because a historical India had enough genuine scientific accomplishment so as not to be debased by absurd exaggerations. It is unlikely that the ISRO chief felt the need to make the claim about a "98 percent" success in the Chandrayaan 2 mission under the influence of the current mood but we would not be surprised if that were the case.

Even at the time writing this book in the third week of October, there were news reports coming out of India about Hindu priests performing Havans or fire worship rituals through various sacrifices to make the Vikram lander miraculously revive.

While as fringe developments these can have some passing amusement value, oftentimes they become so deeply embedded into the popular psyche that they overwhelm the scientific temper even among those who practice science daily. It is true that perhaps unlike any other major country, India's continuous civilization stretching at least some five millennia has blurred the lines between science and faith even though the former has always been an intrinsic part of its makeup.

It is in this context that the overarching political narrative can prove detrimental to the pursuit of science. There is nothing inherently wrong if individual scientists maintain their own faith even while practicing their science but as institutions, the scientific establishment must not give any quarter to it. Rocket science, which is at the heart of any space enterprise, necessitates that precise calculations go into it and not faith. Vikram's apparent crash is a direct consequence of the physics of falling objects. Nothing else can explain that.

Chapter 5

THE SCIENCE OF FALLING OBJECTS

Galileo Galilei

Recorded curiosity about why objects fall and how they do so is at the very least over 2300 years old. It was the Greek philosopher Aristotle (384 BCE to 324 BCE) who first wondered about it. He eventually theorized that objects fall relative to their mass. Two same-sized objects, according to his theory, but with different densities will fall at different speeds when dropped from the same height at the same time.

He was, of course, wrong.

However, there was a gap of some 1800 years before it was known that Aristotle was wrong. It was the 16th century mathematician and scientist Galileo Galilei (1564–1642) who began overturning a nearly 2000-year-old certitude. He employed an exquisitely simple experiment—took a heavy object and a light object and dropped them at the same time from the same height to see which fell the fastest. Air resistance would necessitate that the heavy object hit the ground before the light one. However, to understand how pure gravity works, air has to be removed. In other words, create a perfect vacuum. In a perfect vacuum, a bowling ball and feathers would fall at precisely the same rate and hit at the same time. There have been popular experiments conducted by scientists to illustrate this. One by the physicist Brian Cox at NASA's Space Simulation Chamber located at the Space Power Facility in Ohio was among the most remarkable. With a volume of 22,653 cubic meters, the facility is the largest vacuum chamber in the world.

The red bowling ball and white feathers fell and hit the ground in perfect unison proving Aristotle wrong the way Galileo had said.

As an aside, it is worth mentioning that it was Galileo who first detected mountains on the Moon using a telescope that he himself built. He also viewed the lunar surface as otherwise smooth from a distance.

The reason why we refer to this is because the Moon has no air. It has an extremely thin layer of gases called exosphere. A NASA backgrounder explains, "Until recently, most everyone accepted the conventional wisdom that the moon has virtually no atmosphere. Just as the discovery of water on the moon transformed our textbook knowledge of Earth's nearest celestial neighbor, recent studies confirm that our moon does indeed have an atmosphere consisting of some unusual gases, including sodium and potassium, which are not found in the atmospheres of Earth, Mars or Venus. It's an infinitesimal amount of air when compared to Earth's atmosphere. At sea level on Earth, we breathe in an atmosphere where each cubic centimeter contains 10,000,000,000,000,000,000 molecules; by comparison, the lunar atmosphere has less than 1,000,000 molecules in the same volume. That still sounds like a lot, but it is what we consider to be a very good vacuum on Earth. In fact, the density of the atmosphere at the Moon's surface is comparable to the density of the outermost fringes of Earth's atmosphere where the International Space Station orbits."

It adds, "One of the critical differences between the atmospheres of Earth and the moon is how atmospheric molecules move. Here in the dense atmosphere at the surface of Earth, the molecules' motion is dominated by collisions between the molecules. However, the moon's atmosphere is so thin, atoms and molecules almost never collide. Instead, they are free to follow arcing paths determined by the energy they received from the processes described above and by the gravitational pull of the moon.

The technical name for this type of thin, collision-free atmosphere that extends all the way down to the ground is a "surface boundary exosphere." Scientists believe this may be the most common type of atmosphere in the solar system. In addition to the moon, Mercury, the larger asteroids, a number of the moons of the giant planets and even some of the distant Kuiper belt objects out beyond the orbit of Neptune, all may have surface boundary exospheres. But in spite of how common this type of atmosphere is, we know very little about it. Having one right next door on our moon provides us with an outstanding opportunity to improve our understanding."

Since the Moon has what on Earth would be a very good vacuum, objects landing or falling will have next to no resistance. For all practical purposes, it would be free fall unless it is an artificial landing object such as Vikram, which had boosters that would moderate and control its descent.

One of the key features of studying what might have happened to Vikram is the coefficient of restitution (COR). It is the ratio of relative velocity after a crash to relative velocity just before the crash. Since the Moon's velocity is zero, it becomes a ratio of velocity of Vikram after the crash to velocity of Vikram just before the crash. It relates to a collision between two objects. It is derived as the ratio of the relative velocity after to the relative velocity before the collision. That number ranges between 0 and 1, where 0 is a perfectly inelastic collision and 1 is elastic collision. In lay terms, an elastic collision is like a head-on collision between two objects where they practically merge into each other and splinter. An elastic collision would involve a great deal of bouncing of one or both objects.

The COR depends considerably on what materials the objects are made of. Then there are other determining factors such as the impact velocity, the shape and size of the objects, their location and their temperatures.

Prof. Thakkar, one of the two authors, has for decades studied issues such as the COR. Although his studies of crashes from the standpoint of the COR have all been Earth-bound, the forces that act upon them are universal. That is the beauty of physics. So far most major theories of physics and the mathematics attached to them have been shown to work across the universe. So to that extent a truck hitting a railroad bridge in Ohio in 2004 that Dr. Thakkar studied would in retrospect offer him clues on how Vikram may have hit the lunar surface, with what force and its potential for damage.

A semi traveling at 40 miles an hour suffered this damage after hitting a railroad bridge.(Image courtesy: Bharat Thakkar)

The truck in question was a semi that belonged to AT&T. It carried telecom equipment worth $1.2 million. For a semi that was reportedly going at 40 miles an hour on Earth with all its resistances and frictions, its damage was quite remarkable. (See photos.) Made of steel it took quite a beating externally. More importantly, the expensive equipment inside was severely damaged and flung around violently with some parts smashed beyond repairs. This is as garden variety a collision as it gets. A semi can weigh anywhere between 20,000 lb (9,100 kg) on a single axle, 34,000 lb (15,000 kg) on a tandem, and 80,000 lb (36,000 kg) total for any vehicle or combination. For an object of that size to crash into a stationary railroad bridge and still cause the kind of damage it did, should illuminate what might have happened to Vikram.

It was a measure of the damage that the trucking company insurance had to pay $750,000 to AT& T for its switch damage claim.

Projectile motion comes in many sports, such as pitching in baseball, field goal in football, billiard balls hitting each other, etc. The fire extinguishing trucks throw water jet in a form of a projectile, the same as in a water fountain. According to the Encyclopedia Britannica, it was Galileo who was said to have first "realized that the curved path followed by a missile or projectile is a parabola. He had arrived at his conclusion by realizing that a body undergoing ballistic motion executes, quite independently, the motion of a freely falling body in the vertical direction and inertial motion in the horizontal direction. These considerations, and terms such as ballistic and projectile, apply to a body that, once launched, is acted upon by no force other than Earth's gravity."

It adds, "Projectile motion may be thought of as an example of motion in space—that is to say, of three-dimensional motion rather than motion along a line, or one-dimensional motion. In a suitably defined system of Cartesian coordinates, the position of the projectile at any instant may be specified by giving the values of its three coordinates, $x(t)$, $y(t)$, and $z(t)$. By generally accepted convention, z(t) is used to describe the vertical direction."

The reason why we are citing these rather trivial sport examples is because the COR calculations that go into it are the same as what we have used to calculate Vikram's most likely crash and destruction. It is opportune to raise here an important question whether it was the hardware damage on its crashing that killed Vikram software or it was a glitch on the software that caused the crash and caused the destruction. The reason for mentioning this here is because even a much bigger telecom switch encased in a much bigger semi-trailer suffered so much damage at about 40 miles an hour on Earth with all its moderating atmosphere. That might help us reasonably understand what might have happened to Vikram.

As Dr. Thakkar watched the live broadcast of the last phase of Vikram soft-landing and its subsequent crash, he instantly began to think in terms of factors such as the COR, mechanics, projectile motion, horizontal and vertical velocities and how those would have affected Vikram on the Moon with nearly no atmosphere. Among the first thoughts was whether the systems failure in the boosters meant it was for all practical purposes in a free fall in the last stage of its descent.

These are important questions that require careful examination shorn of emotions and any nationalistic considerations.

> *The lander likely crashed at about 184 miles per hour, a velocity at which nothing of value could have survived.*

Chhaya's immediate intuitive deductions of Vikram's failure even before it was acknowledged officially led him to put an update on his Facebook timeline indicating that the soft-landing aspect of the mission appeared to have failed. He was particularly struck by the sudden deviation after the rough braking phase that appeared on the telemetry graph on the screen. Admittedly, it was a non-scientific response triggered by a combination of what was on display in the lander's telemetry and intuition. Interestingly, within minutes of the wobble in the telemetry that Chhaya noticed during the live ISRO broadcast, he came upon a tweet by Cees Bassa, an astronomer at the Netherlands Institute for Radio Astronomy on September 6.

Cees Bassa
@cgbassa

Replying to @cgbassa @radiotelescoop and 4 others

It looks like the @isro #Chandrayaan2 Vikram lander has crashed. After the rough braking phase the Doppler curve from @radiotelescoop shows some wiggles, and then, at 20:20:01UTC the signals disappeared...

3:26 PM · Sep 6, 2019 · Twitter Web App

632 Retweets 1.1K Likes

Bassa said, "It looks like the @isro #Chandrayaan2 Vikram lander has crashed. After the rough braking phase, the Doppler curve from @radiotelescoop shows some wiggles, and then, at 20:20:01UTC the signals disappeared..."

Dr. Thakkar, meanwhile, got on with the actual mathematics involved. He quickly established that the lander likely crashed at about 184 miles per hour, a velocity at which nothing of value could have survived. On the basis of the points that he sent me I reported a news story for the IANS wire, India's largest independent news service, saying precisely that. That news story was published widely in the India media and created quite a bit of a debate.

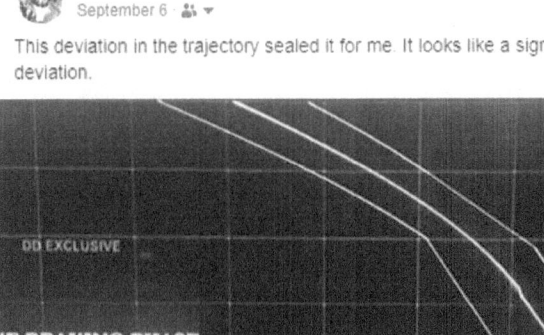

Several fundamental questions came to Dr. Thakkar's mind about the lander. These questions may be uncomfortable to those lay observers not trained in mechanics, projectile motion, horizontal and vertical velocities etc. They certainly go beyond any politics and ideology. Two in particular are worth mentioning here.

Vikram was designed for soft landing.

• Is it true that there was no contingency in design for free fall any time during moon landing or any other time during Chandrayaan-2 mission?

- Expecting Vikram to function after free fall is a false hope similar to a 'dead person' coming alive. Pulling the plug on 'brain dead' is the right attitude. Expecting for divine intervention is very much cultural phenomenon in India.

Nearly three days after the crash, ISRO seemed optimistic that they may be able to establish contact with it. In a tweet on September 9 it announced that the lander had been "located by the orbiter of Chandrayaan 2 but no communication yet. All possible efforts are being made to establish communication with (the) lander."

Perhaps reflecting the irrational optimism pervading India's air in the immediate aftermath of the crash, media reports quoted an unnamed official to say that the lander was intact but ISRO offered no official response.

On September 10, Kenneth Chang of The New York Times quoted Mark Johnson, the lead for the Lockheed Martin team during the successful landing of NASA's InSight spacecraft on Mars in November, as saying, "It's hard enough to land on another planetary body when everything works perfectly but to have done it in the presence of a major anomaly would have been astounding."

While that turned out to be a false hope, one of the fundamental questions that troubled Dr. Thakkar was whether there was consideration in Vikram's design about "the moon's 'coefficient of restitution' at the predetermined landing location."

> *It may be reasonable to surmise that only the first hit may have crushed Vikram. There could have been no chance for subsequent bouncing. However, theoretically it is possible to have more bouncing like a tennis ball but Vikram is not a tennis ball.*

Upon free falling (from 2.1 km height), assuming the COR of 0.5, Vikram would have bounced back and forth more than ten times in an upright position, which would be positive thinking, before coming to rest at a different location with a different coefficient of restitution than originally planned. During bouncing back and forth, the pre-determined center of gravity of Vikram could have gone into non-equilibrium position upon the first impact. Thus, is it not hopeless to think for Vikram to stay in upright position?

A post-mortem must be carried out in terms of the mechanical design of Vikram. What was the factor of safety used in the dynamic mechanical design of Vikram? Was it even considered? These are the questions that need to be asked to ensure that upcoming missions do not suffer a similar fate.

Arriving at the figure of 184 miles an hour crash was a straightforward calculation. Free fall velocity can be closer to square root of 2gh. Where g is one-sixth of Earth (32.2 feet per Sec2) h = 2.1 km = approximately 7000 feet. With this free fall velocity is 270 ft per second = 184 miles per hour, nothing can survive at that velocity.

We are assuming 'free fall' but it is not exactly so. It is a minor thing but needs to be mentioned. At 2.1 km height, because of the existing trajectory, there must be initial downward velocity. This is equivalent to forcing the tennis ball downward with a velocity that can rebound greater than 'free fall'. This downward velocity and coefficient of restitution combined can give us the worse first bounce than mentioned earlier.

It may be reasonable to surmise that only the first hit may have crushed Vikram. There could have been no chance for subsequent bouncing. However, theoretically it is possible to have more bouncing like tennis ball but Vikram is not a tennis ball. We also have to think in terms of the whether the location where it hit was rocky, somewhat rocky or fully powdery like sand on Earth. That detail would have had discernible impact on whether it bounced or just crashed once and settled down significantly broken. This is where the COR between 0 and 1 comes into play. For the sake of argument, let's say it was 0.5.

If COR is 0.5, the first bounce would be 0.525 km and the next bounce will be 0.13 km and so on. The formula will be square root of ratio height after rebound to height for next rebound.

Anything four-legged falling does not necessarily hit all four legs at the same time depending upon the profile of the Moon's surface. It is likely one leg touching first and taking all the forces due to free falling. If one leg is weakened, the rest cannot help the situation. The deformations that are plastic are not recoverable in metal and materials. Anyone making prediction with a common sense, Vikram was likely destroyed as soon as it hit the moon ground.

Dr. Thakkar's calculation of 184 miles an hour was based on his immediate calculation and has not necessarily been disputed. However, in his report of September 10, Chang of The New York Times mentioned efforts by amateur astronomers who used a 25-meter radio telescope belonging to the Netherlands Institute for Radio Astronomy. Quite fortuitously, he quoted the same Cees Bassa to report that the astronomer had calculated the velocity of the lander using the Doppler data to be more than 110 miles an hour. Whether it was 184 miles an hour going by pure mathematics or more than 110 miles using the Doppler data, its crash at that velocity was unlikely to have kept it intact.

Cees Bassa
@cgbassa

Replying to @cgbassa and @ProjectJupyter

One conclusion from this analysis is that it is now very likely that the #Chandrayaan2 signal was lost on impact with the surface, not at 2 km altitude as initially reported. That does not bode well for attempts to regain contact, as it suggests the lander was damaged on impact.

4:11 AM · Sep 10, 2019 · Twitter Web App

The wobble during the last minutes of the descent, as Chhaya described it, or "oscillations" as Bassa and other called them using a more technical term they essentially speak of the same occurrence.

One of the important questions that Dr. Thakkar raised with Chhaya both by email and phone was it was the hardware damage on its crashing that killed Vikram software or it was a glitch on the software that caused the crash and caused the physical destruction. Dr. Thakkar had already suggested that at the velocity he worked out could not have spared the lander physically. In other words, if it broke on hitting the ground it could not have sent a signal after that.

Bassa had tweeted on September 10, "One conclusion from this analysis is that it is now very likely that the #Chandrayaan 2 signal was lost on impact with the surface, not at 2 km altitude as initially reported. That does not bode well for attempts to regain contact, as it suggests the lander was damaged on impact."

This convergence of views was interesting.

We are sure ISRO would have examined a particular aspect in the Vikram lander failure that was discovered in an Israeli mission. As The New York Times reported on September 6, 2019 in a story datelined Bangalore By Jeffrey Gettleman, Kenneth Chang, Kai Schultz and Hari Kumar, "An Israeli nonprofit sent a small robotic spacecraft named Beresheet to the moon, but its landing attempt in April went awry in a manner similar to Chandrayaan-2. The initial descent went as planned, but then communications were lost near the surface. It was later discovered that a command to shut off the engine was incorrectly sent."

Was a command to shut off the engine incorrectly sent? That may be one of the questions. Over a month and half after the mission failure and at the time of writing this book, the authors had not come across any reports whether ISRO had made an initial assessment, including the kind of problem that Beresheet faced, and made it public.

Although somewhat tangential, it might be useful to mention here the matter of the relatively low budget for both the Chanrayaan 2 and Beresheet missions compared to what NSA and ESA would typically spend. In some ways it has been a matter of some national pride that India manages its missions so frugally (between $100 and $150 million in both cases). There may be case that can be made that excessive budget constraints can escalate some risks to such missions even though it is hard to specifically identify how that might be the case.

Vikram's failure understandably set off a mixture of emotions across India. While ISRO tried to put its best spin on it as evident in its chief Sivan's comments, it is safe to think that internally they were aware of the seriousness of the failure. NASA had the reaction worthy of an agency, which has had its own spectacular successes as well as dramatic failure even as on balance it has remained the leading space enterprise on the planet. In a tweet on September 7, it said, "Space is hard. We commend @ISRO's attempt to land their #Chandrayaan2 Mission on the Moon's South Pole. You have inspired us with your journey and look forward to future opportunities to explore our solar system together."

Space is indeed hard particularly when it comes to soft-landing spacecraft anywhere because it requires precise navigation that is at the mercy of the onboard computer.

Originally designed to work for 14 days or one lunar day, the Vikram lander remained untraceable even after over one and half months after its assumed crash on September 7. Hopes of finding Vikram appeared to be fading by October 23 as reported by the Press Trust of India in the following story:

NASA has found no evidence of Chandrayaan 2's Vikram lander in the images captured during a latest flyby of its Moon orbiter of the lunar region where India's ambitious mission attempted a soft landing, the US space agency said.

On 7 September, Indian Space Research Organisation (ISRO) attempted a soft landing of Vikram on the uncharted lunar south pole, before losing communication with the lander.

"The Lunar Reconnaissance Orbiter imaged the area of the targeted Chandrayaan 2 Vikram landing site on 14 October but did not observe any evidence of the lander," Noah Edward Petro, the Project Scientist for the LRO mission, told PTI in an exclusive email interaction.

Petro said that the camera team carefully examined the images and employed the change detection technique — using a ratio of an image from prior to the landing attempt to the one acquired on 14 October.

This approach, he said, is used for finding new meteorite impacts on the Moon that also helped locate the recent Beresheet lander.

"It is possible that Vikram is located in a shadow or outside of the search area. Because of the low latitude, approximately 70 degrees south, the area is never completely free of shadows, John Keller, Deputy Project Scientist LRO Mission, told PTI.

During its previous flyby, the LRO passed over the landing site on September 17 and acquired a set of high-resolution images of the area.

Although one failed mission does not and will not create a terminal setback for ISRO, it will certainly force them to closely examine every aspect of that failure in order not to repeat it. The mission was on what might be considered a shoestring budget, but it is still a considerable amount of public money. It is ISRO's responsibility to offer the nation a detailed postmortem report as well as a lessons-learned report. Ordinary citizens may not necessarily understand them in all their complex details, but they would at least have the consciousness that even when a great enterprise such as

ISRO fails it feels the responsibility to account for itself. Sentimentalism and pride have a place in a nation's life but not in the annals of serious scientific analysis when a failure of this kind occurs.

Since such missions involve large sums of public money, ISRO and those involved must ask difficult questions of them beyond just "sentimentalism". ISRO is a great enterprise and has done India proud. It has within its DNA the ability to ask brutally honest questions and come up with rational answers even if they run counter to the prevailing popular political rhetoric.

The ISRO did set up a committee consisting of academics and agency experts to analyze the cause of the lander's failure and the subsequent loss of communication. It was said by an official that the committee was likely to submit its report in about a month, a deadline that had already passed at the time of the publication of this book.

Space missions are a product of centuries of intensive scientific theories, research and calculations about subjects as diverse as gravity, motion, trajectory, projectile, rocketry, metallurgy, telemetry, communications and so on. The variables are so numerous in each of these categories that it is remarkable that they have all come together so successful for decades because of the single-minded focus of scientists and engineers among space-faring nations, including India.

Chapter 6

MOON AND ITS UNEVEN GRAVITY

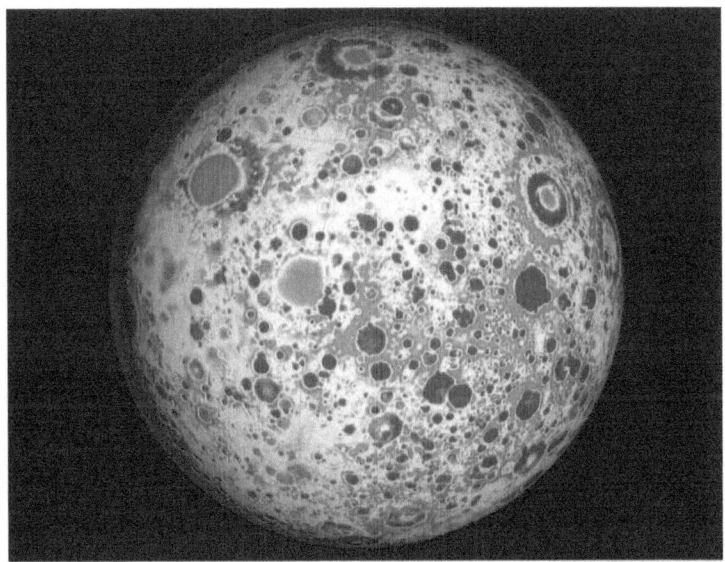

This image shows the variations in the lunar gravity field as measured by NASA's Gravity Recovery and Interior Laboratory (GRAIL) during the primary mapping mission from March to May 2012. Image Credit: NASA/JPL-Caltech/MIT/GSFC

For long, the Moon's gravity had been found to be mysteriously uneven. It was also significantly uneven to have a decisive impact on spacecraft attempting to land. As early as 1968, scientists at the Jet Propulsion Laboratory (JPL) were mystified at this lunar mystery. That the uneven lunar gravity could potentially throw a landing spacecraft off course was a real possibility.

It is not a subject that has been mentioned in the context of the Vikram failure. It is not our case that that may have played any role at all but as a reference for any upcoming missions to the Moon, we think it might be interesting to bear in mind.

NASA's Gravity Recovery and Interior Laboratory (GRAIL) mission has uncovered the origin of massive invisible regions that make the moon's gravity uneven, a phenomenon that affects the operations of lunar-orbiting spacecraft.

As pointed out earlier, the uneven gravity is a result of the hidden presence of mascons, a short for a mass concentration, which are essentially a concentration of denser material below the lunar surface which exercise a greater gravitational pull than its normal force. The authors are not aware if ISRO took this factor into consideration or whether it was even relevant to the Vikram landing in the chosen area.

The mystery of the uneven lunar gravity was solved in 2013 when NASA made the following announcement. It is important to carry it in its entirety here because it explains in precise terms what we had alluded to.

"NASA's Gravity Recovery and Interior Laboratory (GRAIL) mission has uncovered the origin of massive invisible regions that make the moon's gravity uneven, a phenomenon that affects the operations of lunar-orbiting spacecraft.

Because of GRAIL's findings, spacecraft on missions to other celestial bodies can navigate with greater precision in the future.

GRAIL's twin spacecraft studied the internal structure and composition of the moon in unprecedented detail for nine months. They pinpointed the locations of large, dense regions called mass concentrations, or mascons, which are characterized by strong gravitational pull. Mascons lurk beneath the lunar surface and cannot be seen by normal optical cameras.

GRAIL scientists found the mascons by combining the gravity data from GRAIL with sophisticated computer models of large asteroid impacts and known detail about the geologic evolution of the impact craters. The findings are published in the May 30 edition of the journal Science.

"GRAIL data confirm that lunar mascons were generated when large asteroids or comets impacted the ancient moon, when its interior was much hotter than it is now," said Jay Melosh, a GRAIL co-investigator at Purdue University in West Lafayette, Ind., and lead author of the paper. "We believe the data from GRAIL show how the moon's light crust and dense mantle combined with the shock of a large impact to create the distinctive pattern of density anomalies that we recognize as mascons."

The origin of lunar mascons has been a mystery in planetary science since their discovery in 1968 by a team at NASA's Jet Propulsion Laboratory in Pasadena, Calif. Researchers generally agree mascons resulted from ancient impacts billions of years ago. It was not clear until now how much of the unseen excess mass resulted from lava filling the crater or iron-rich mantle upwelling to the crust.

On a map of the moon's gravity field, a mascon appears in a target pattern. The bulls-eye has a gravity surplus. It is surrounded by a ring with a gravity deficit. A ring with a gravity surplus surrounds the bulls-eye and the inner ring. This pattern arises as a natural consequence of crater excavation, collapse and cooling following an impact. The increase in density and gravitational pull at a mascon's bulls-eye is caused by lunar material melted from the heat of a long-ago asteroid impact.

"Knowing about mascons means we finally are beginning to understand the geologic consequences of large impacts," Melosh said. "Our planet suffered similar impacts in its distant past, and understanding mascons may teach us more about the ancient Earth, perhaps about how plate tectonics got started and what created the first ore deposits."

This new understanding of lunar mascons also is expected to influence knowledge of planetary geology well beyond that of Earth and our nearest celestial neighbor.

"Mascons also have been identified in association with impact basins on Mars and Mercury," said GRAIL principal investigator Maria Zuber of the Massachusetts Institute of Technology in Cambridge. "Understanding them on the moon tells us how the largest impacts modified early planetary crusts."

Launched as GRAIL A and GRAIL B in September 2011, the probes, renamed Ebb and Flow, operated in a nearly circular orbit near the poles of the moon at an altitude of about 34 miles (55 kilometers) until their mission ended in December 2012. The distance between the twin probes changed slightly as they flew over areas of greater and lesser gravity caused by visible features, such as mountains and craters, and by masses hidden beneath the lunar surface."

It would be interesting to find out if during the planning of the soft landing the Vikram team was aware of and had considered this contingency at all. To the extent that any descending object, no matter how big or small and how under controlled circumstances such as Vikram would be impacted by gravity directly, the lunar gravity unevenness would logically be kept in view while planning the soft-landing. It is conceivable that the effect of the uneven lunar gravity, if at it was in the region where Vikram was to land, was so minuscule as to be irrelevant. However, we as authors do not know that from a distance and hence feel the need to raise the possibility.

Chapter 7

BEYOND THE VIKRAM CRASH

Two ISRO scientists in moments of realization of the possible crash. (Image: ISRO broadcast)

All great scientific establishments have resilience built into them. ISRO is no exception.

Although the prospects of Vikram being alive were slim to none despite expectations of some divine intervention brought forth by fire worship rituals and other such actions, ISRO had to get on with the task of its spacecraft carrying out its functions normally and some even unexpectedly beyond.

One of the tasks for the spacecraft orbiting around the Moon is to study various features of its surface. Since its formation the Moon has been under constant assaults from meteorites, asteroids and comets creating a vast number of impact craters. As ISRO explains, "Impact craters are approximately circular depressions on the surface of the Moon, ranging from small, simple, bowl-shaped depressions to large, complex, multi-ringed impact basins. In contrast to volcanic craters, which result from explosion or internal collapse, impact craters typically have raised rims and floors that are lower in elevation than the surrounding terrain. The study of the nature, size, distribution and composition of impact craters and associated ejecta features reveal valuable information about the origin and evolution of craters. Weathering processes result in many of the crater physical features and ejecta material get covered by layers of regolith, making some of them undetectable using optical cameras. Synthetic Aperture Radar (SAR) is a powerful remote sensing instrument for studying planetary surfaces and subsurface due to the ability of the radar signal to penetrate the surface. It is also sensitive to the roughness, structure and composition of the surface material and the buried terrain."

"Previous lunar-orbiting SAR systems such as the S-band hybrid-polarimetric SAR on ISRO's Chandrayaan-1 and the S & X-band hybrid-polarimetric SAR on NASA's LRO, provided valuable data on the scattering characterisation of ejecta materials of lunar impact craters. However, L & S band SAR on Chandraayan-2 is designed to produce greater details about the morphology and ejecta materials of impact craters due to its ability of imaging with higher resolution (2 - 75m slant range) and full-polarimetric modes in standalone as well as joint modes in S and L-band with wide range of incidence angle coverage (9.5° - 35°). In addition, the greater depth of penetration of L-band (3-5 meters) enables probing the buried terrain at greater depths. The L & S band SAR payload helps in unambiguously identifying and quantitatively estimating the lunar polar water-ice in permanently shadowed regions."

"Imaging Infrared Spectrometer (IIRS) on-board Chandrayaan-2 is designed to measure the reflected sunlight and emitted part of Moon light from the lunar surface in narrow and contiguous spectral channels (bands) ranging from ~800 - 5000 nanometer (0.8-5.0 micrometer (µm)). It uses a grating to split and disperse the reflected sunlight (and emitted component) into different spectral bands. The major objective of IIRS is to understand the origin and evolution of the Moon in a geologic context by mapping the lunar surface mineral and volatile composition using signatures in the reflected solar spectrum."

As it frequently happens space missions have a way of unexpectedly producing insights they were not necessarily designed for. It happened with the Chandrayaan 2 spacecraft on September 30 and October

1. It captured the measurements of some solar flares. It became possible because of an instrument called Solar X-ray Monitor which, along with Chandrayaan 2 Large Area Soft X-ray Spectrometer or CLASS is mainly supposed to help "to measure the lunar elemental composition."

ISRO explained, "Many violent phenomena continuously keep occurring on surface of the Sun and its atmosphere known as the corona. This solar activity follows an eleven-year cycle, which means, it goes through its 'solar maxima' and 'solar minima' once every eleven years. While the cumulative emission of solar X-rays emitted over a year varies with the solar cycle, these are often punctuated with extremely large x-ray intensity variations over very short periods, few minutes to hours. Such episodes are known as solar flares.

Chandrayaan-2 orbiter utilizes X-rays emitted by the Sun in a clever way to study elements on the lunar surface. Solar X-rays excite atoms of constituent elements on the lunar surface. These atoms when de-excited emit their characteristic X-rays (a fingerprint of each atom). By detecting these characteristic X-rays, it becomes possible to identify various major elements of the lunar surface. However, in order to determine their concentration, it is essential to have simultaneous knowledge of the incident solar X-ray spectrum.

The Chandrayaan 2 orbiter carries two instruments, Chandrayaan 2 Large Area Soft X-ray Spectrometer (CLASS) and Solar X-ray Monitor (XSM), to measure the lunar elemental composition using this technique. Here, the CLASS payload detects the characteristic lines from the lunar surface and the XSM payload simultaneously measures the solar X-ray spectrum."

The advantage with Chandrayaan-2's Solar X-ray Monitor is that it is more sensitive than the other currently deployed solar flare monitors. Such unexpected spinoffs can often turn an average mission into something a little more special.

Had Vikram landed successfully, the rover within it, Pragyan would have moved about on the lunar surface on its six wheels at one centimeter per second, eventually covering half a kilometer. The rover was equipped with a Laser Induced Breakdown Spectroscope (LIBS) that would have helped identify the elements around the landing site.

Other than that, the Orbiter High resolution Camera acquired very high spatial resolution images of the Moon from an orbit of 100 km. They are said to be the sharpest images ever by a lunar orbiter platform. As is the case with most partially successful space missions, Chandrayaan-2 too had produced and will continue to produce some important insights into the Moon and sometimes solar flares.

Had Vikram landed successfully, the rover within it, Pragyan would have moved about on the lunar surface on its six wheels at one centimeter per second, eventually covering half a kilometer. The rover was equipped with a Laser Induced Breakdown Spectroscope (LIBS) that would have helped identify the elements around the landing site. It also carried an Alpha Particle Induced X-ray Spectroscope (APIXS) designed to study the composition of various elements in the lunar surface.

These were potentially important tasks which are now rendered useless because of the crash. The extent of loss in scientific research is hard to quantify because what the rover might have found is imponderable at this stage. However notional it may seem, it is a loss nonetheless.

Chapter 8

RELIABILITY AND QUALITY CONTROL

One heartening sign of the Vikram enterprise was the presence of school children inside the mission control watching the events unfold in real time. At a time when the political rhetoric is so jumbled up with often non-sensical stories of India's ancient mastery over all aspects of science, some pushed by no less a personage than the prime minister himself, it is important that children are exposed to the rough and tumble of science.

As India seeks a higher and inclusive economic growth for all its citizens, the role of science and technology cannot be overemphasized. India has a long history ordinary citizen deploying their individually crafted solutions to their everyday problems. It goes by the umbrella term Jugaad which is essentially a quick fix to problems that the elite scientific and technological establishment of the country finds beneath them to address. Jugaad is often not replicable and necessarily crude in its design and execution simply because it is a quick fix put together as a last resort.

While Jugaad may work for individuals, for a nation aspiring to become a $5 trillion economy by 2025 by nearly doubling the current size of $2.7 billion it needs a gigantic attitudinal transformation. It is in this context that it is important that schools across the country expose young children to all manners of science as part of their daily curriculum. Given the lure and mystery of space, it is a great starting point but it is essential that school education becomes so comprehensive as to include all walks of life.

The Science channel here in America has a remarkable series called 'How It's Made' which offers an astonishing range of how anything is made. It has been on since 2001 and is a rich resource to excite young minds. Watch any episode of the series randomly and one begins to get a measure of how much science, technology and engineering go into making even the smallest of things from paperclips and rubber bands to giant parts of very high-end sophisticated machinery.

We believe it is essential that the next generation of scientists, engineers and entrepreneurs in India are equipped with such knowledge in order to grapple with challenges of the 21st century and beyond. At a time when the world is on the threshold of quantum computing as teased by Google's quantum supremacy, it is imperative that a country like India with such a wide range of raw scientific and inventive talent does not remain mired in just Jugaad.

Inextricably linked to the pursuit of overall excellence in science, technology and engineering, or for that matter any aspect of life, is strict adherence to reliability of systems and quality control. Dr. Thakkar has spent close to five decades on those particular themes in America as well as in India apart from other parts of the world. Quite apart from approaching the Vikram lander problem from a mathematical standpoint and arriving at specific velocity calculations, his overarching concern has been about how India looks at reliability and quality control of systems.

Over the years, Dr. Thakkar and Chhaya have had many conversations about these themes. One of the themes in particular has been what is described as a complex system. Its most obvious examples are commercial aircraft, spacecraft, rockets, submarines, naval aircraft carriers, Internet, and even landers and rovers. All of these systems require reliability and quality control.

The International Council on Systems Engineering (INCOSE) is a not-for-profit membership organization founded "to develop and disseminate the interdisciplinary principles and practices that enable the realization of successful systems. INCOSE is designed to connect SE professionals with educational, networking, and career-advancement opportunities in the interest of developing the global community of systems engineers and systems approaches to problems." It is regarded as a key group that advocates a systems approach. While the body was founded in 1990, as a much-respected engineer, Dr. Thakkar has for decades put in practice much of what the INCOSE has recommended.

The INCOSE lists the following reasons why projects fail:

1. Incomplete requirements   13.1%

2. Lack of user involvement   12.4%

3. Lack of resources   10.6%

4. Unrealistic expectations   9.9%

5. Lack of executive support   9.3%

6. Changing requirements/specs   8.7%

7. Lack of planning   8.1%

8. Didn't need it any longer   7.5%

Any failure such as the Vikram landing has one overriding question attached to it: What went wrong? We have not necessarily discussed what may have gone wrong with Vikram because that is not out core expertise. Instead on the basis of mathematical calculations we have surmised whether Vikram could have survived the crash.

However, in India's quest to become a $5 trillion economy by 2025, which means growing by nearly $500 million in GDP every year until then, it will have to radically alter its core attitudes from just Jugaad to a massive transformation via reliability and quality control. It is not Dr. Thakkar's case that reliability engineering is the solution to all the world's systems problems but it will certainly help manage some of them.

Apple's much celebrated catchphrase "Think differently" is not necessarily that different because it has been suggested in its various forms for a very long time. That includes famously by the poet Robert Frost in 'Road not taken" where he says,

> *Two roads diverged in a wood, and I,*
> *I took the one less traveled by,*
> *And that has made all the difference.*

Even thinking differently does not preclude thinking of reliability of systems. One of Dr. Thakkar's favorite examples is to say:

Every morning you better be running...

"Every morning in Africa, a gazelle wakes up. It knows it must run faster than the fastest lion or it will be killed...

Every morning a lion wakes up. It knows it must outrun the slowest gazelle or it will starve to death.

It doesn't matter whether you are a lion or a gazelle...or an employee when the sun comes up, you'd better be running."

On the face of it, it seems pretty straightforward when it comes to reliability. One starts with the awareness that all products require some sort of reliability. Reliability brings profit, which can mean both in pecuniary sense for a business or a technological success for an enterprise like ISRO.

Doing anything without paying attention to reliability is like sailing in an ocean without direction. It has to be designed into the product.

> *In many ways, Vikram can become a transformational example for India not just in space science but across the entire spectrum of nation building. It must start at the level of curiosity among children about how things work and how they are made.*

In many ways, Vikram can become a transformational example for India not just in space science but across the entire spectrum of nation building. It must start at the level of curiosity among children about how things work and how they are made. There is an urgent need for content producers in India to produce series along the lines of 'How It's Made'. To begin with perhaps the already existing library of content with the channels here in America could be acquired for nationwide school distribution. That can be followed up with Indian producers creating such series to suit India's conditions.

We emphasize curiosity because that is where everything begins in life. Unless India's children are encouraged to be curious, empowered to be questioning of everything and then equipped to be able to find answers, there cannot be the kind of economic leaps that the country's political class tries to sell.

With curiosity as the first step, the next must be about a clear awareness of why reliability and quality are absolutely essential. We are not sure if after the Vikram failure any ISRO scientist or teachers who might have accompanied them, bothered to explained to those select few students what really went wrong. Even without getting into the complex details of what might have gone wrong, it would have been useful to explain to them possible causes.

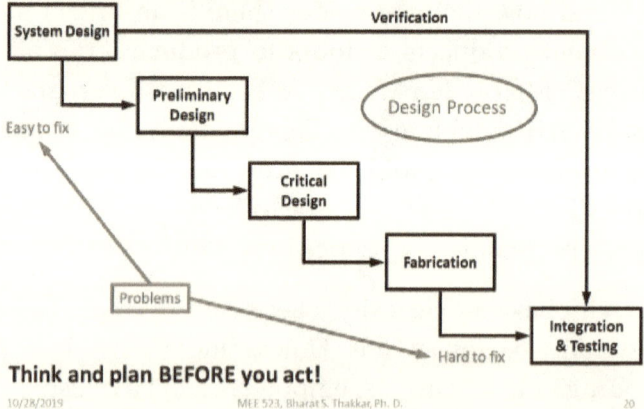

Epilogue

As of publishing this short primer of a book in the last week of October, we had not come across an official ISRO announcement of the findings of the committee of consisting of academics and agency experts to analyze the cause of the lander's failure and the subsequent loss of communication.

The partial mission failure has also understandably receded from the public's mind after the first couple of days of angst over it in the media as well as social media. That is, of course in keeping with any development in life that has a collective interest and attention. There is nothing unusual about it.

The authors were prompted primarily by Dr. Thakkar's calculation about the velocity of the likely crash of Vikram and why most it could not have remained intact after hitting the lunar surface. There has been no development at all in terms of either clearly locating the lander and securing a signal from it as of now. The safe assumption so far that it is lost for now and with the exception of neither NASA nor ISRO has been able to locate the likely wreckage of Vikarm.

In terms of activities subsequent to the mission, ISRO said on October 24 that some 70 participants took part in its 'Chandrayaan-2 data users meet'. The meet "focused on the users who presented their ideas and approach towards data analysis and the outstanding science issue which they plan to tackle using Chandrayaan 2 payload data." These are useful activities related to any such space missions since they expand the pool of expertise of researchers not attached with or employed by ISRO. Such analyses help augment ISRO's core work without any direct financial investment.

Our hope via this easily accessible book is that young Indians excited by the drama of space missions for a few hours of the live broadcast would find it useful to read and get a better handle on the enterprise. Even minor details such as the what the Moon is like in terms of its composition and dimensions are useful to be chronicled as part of this handy book.

Even if just a few students in India and elsewhere in the world feel prompted on reading this book to get involved in the sciences as well as pay greater attention to reliability of systems and quality control, we would feel gratified.

Extra update

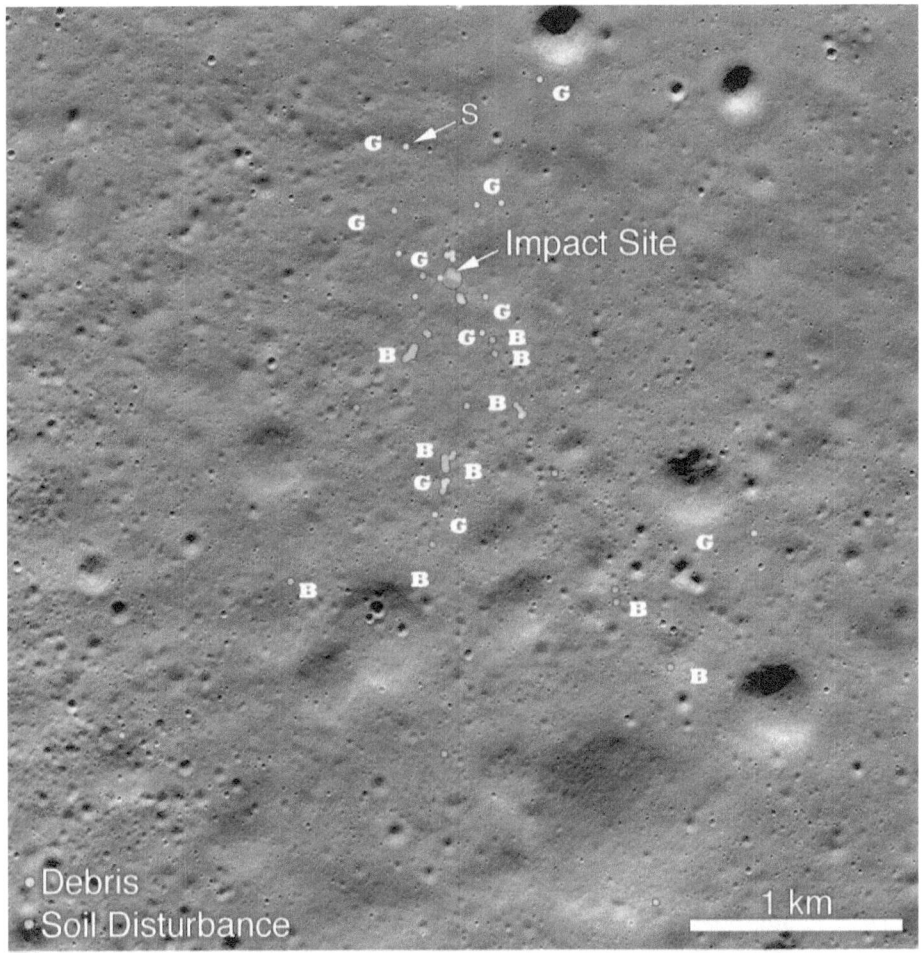

Since the images are printed in black and white we have indicated G for Debris and B for Soil disturbance from NASA's blue and green dots. The image was acquired November 11. Credits: NASA/Goddard/Arizona State University

After the first publication of this slim volume a couple of developments have happened which draw the authors' attention.

One was an official statement in India's parliament by Jitendra Singh, Minister of State in the Prime Minister's Office on November 20, 2019. He was quoted by the Press Trust of India (PTI) wire as saying that "During the second phase of descent, the reduction in velocity (of the lander) was more than the designed value. Due to this deviation the initiation conditions at the state of the fine braking phase were beyond the designed parameters. As a result, Vikram hard-landed within 500 meters of the designated site."

This claim of 500 meters prompted Dr. Thakkar to embark on some fresh calculations. His calculations showed that Vikram would have been off the designated site by more than three times that distance. The figure that Dr. Thakkar worked out was 1680 meters.

The angle of descent was 38 degrees at 2.1 km. ISRO should not try to find the wreckage at 500 meters left of original location.

There is a new element about vertical velocity of 59 km per second, which makes crashing worse than previously thought. Before it was assumed free fall, means at 2.1 km altitude, zero vertical and horizontal velocities.

There is no way the Vikram had a hard landing. It had a crash landing. ISRO is still playing soft ball making the event of little consequence.

1. Distance from original intended location is 1680 meters, not 500 meters.

2. If Y-velocity was 59 meters per second, then crashing velocity is 335 feet per second.

See the math below based upon no resistance of air on moon.

The claim that Vikram crashed 500 meters left of original intended location is wrong. It fell at 1680 meters left of desired location.

Y-velocity was 59 meters per second at 2.1 meters. X-velocity was 49 meters per second. That gives angle Theta as Tan inverse 49 divided by 59, comes to 38 degrees with vertical. ISRO should look for crashed Vikram at 1680 meters left of original Intended location.

Since Y-velocity is 59 meters per second, crashing velocity could be

V (crashing) **2 =(V-y)**2 +(2*g of moon*2100)

=(59**2)+ 2*(9.81divided by 6)(2100)

=3481 + 6867

= 10348

=102 meters per second.

V-crashing = 102 meters per second

= 335 feet per second.

The same for "free fall" would be 83 meters per second, 272 feet per second, which was never a free fall.

In Dr. Thakkar's opinion, this is the correct crashing velocity.

That brings us to the second development as reported by NASA on December 2. Hearteningly that development came because of some greatly diligent work by a Chennai-based engineer Shanmuga Subramanian. NASA said this: *"The Chandrayaan 2 Vikram lander was targeted for a highland smooth plain about 600 kilometers from the south pole; unfortunately the Indian Space Research Organisation (ISRO) lost contact with their lander shortly before the scheduled touchdown (Sept. 7 in India, Sept. 6 in the United States). Despite the loss, getting that close to the surface was an amazing achievement. The Lunar Reconnaissance Orbiter Camera team released the first mosaic (acquired Sept. 17) of the site on Sept. 26 and many people have downloaded the mosaic to search for signs of Vikram. Shanmuga Subramanian contacted the LRO project with a positive identification of debris. After receiving this tip, the LROC team confirmed the identification by comparing before and after images. When the images for the first mosaic were acquired the impact point was poorly illuminated and thus not easily identifiable. Two subsequent image sequences were acquired on Oct. 14 and 15, and Nov. 11. The LROC team scoured the surrounding area in these new mosaics and found the impact site (70.8810°S, 22.7840°E, 834 m elevation) and associated debris field. The November mosaic had the best pixel scale (0.7 meter) and lighting conditions (72° incidence angle).*

The debris first located by Shanmuga is about 750 meters northwest of the main crash site and was a single bright pixel identification in that first mosaic (1.3 meter pixels, 84° incidence angle). The November mosaic shows best the impact crater, ray and extensive debris field. The three largest pieces of debris are each about 2x2 pixels and cast a one pixel shadow."

While the early debris is said to be 750 meters, it is apparent from the extensive nature of the debris field as captured by NASA's LROC that the lander's pieces are scattered quite a bit. Going by the blue and green dots in the NASA image it is obvious that the COR was not zero but between value between zero and one. From the looks of it the location seems rocky and Dr. Thakkar's view is that the COR could be between 0.5 and 0.8 That means that the pieces bounced several times. It should not be surprising that if some of the debris is about the distance Dr. Thakkar pointed out and perhaps even beyond.

As an aside, we are struck by the fact that neither NASA nor ISRO has used the term COR publicly unless, of course, in their internal reviews they have factored that in.

It is particularly interesting for the authors that Dr. Thakkar's initial instant analysis within a couple of days of the crash that the lander would have broken down into several pieces appears to have been borne out by the image taken by the NASA debris field and studied by the young Subramanian.

The new data significantly reinforce the authors' original assessment.

www.ingramcontent.com/pod-product-compliance
Lightning Source LLC
Chambersburg PA
CBHW030947240526
45463CB00016B/2043